# THE BIOMASS SPECTRUM

ॐ

*Complexity in Ecological Systems Series*

# THE BIOMASS SPECTRUM

## A Predator-Prey Theory
of Aquatic Production

S. R. KERR AND L. M. DICKIE

COLUMBIA UNIVERSITY PRESS
NEW YORK

Columbia University Press
Publishers Since 1893
New York      Chichester, West Sussex

Copyright © 2001 Columbia University Press
All rights reserved

Library of Congress Cataloging-in-Publication Data

Kerr, S. R.
The biomass spectrum : a predator-prey theory of aquatic production /
S. R. Kerr and L. M. Dickie
p. cm. — (Complexity in ecological systems series)
Includes bibliographical references (p. ).
ISBN 0–231–08458–7 (cloth : alk. paper) — ISBN 0–231–08459–5 (pbk. : alk. paper)
1. Bioenergetics. 2. Biomass. 3. Predation (Biology) 4. Fishes—Size. 5. Fishery
management.   I. Dickie, Lloyd Merlin, 1926–   II. Title.   III. Series.

QH510 .K45 2001
577.6'13—dc21
00–065961

∞

Casebound editions of Columbia University Press books are printed on permanent
and durable acid-free paper.
Printed in the United States of America
c 10 9 8 7 6 5 4 3 2 1
p 10 9 8 7 6 5 4 3 2 1

In memory of
A. G. Huntsman and F. E. J. Fry,
two mentors who showed us that inquiry buoys the spirit, while
conclusions can be weights for the soul

# Contents

∾

ᐤ

Note that the letters "a," "b," "c," and "d" are used in chapter 3 as general empirical constants found in the fitting of data to analytical equations and in discussing their values in the particular context of the fittings. Interpretations of constants are used to assign particular fixed symbols to them. Relationships among a number of the symbols are schematically defined in figure 4.1.

Fixed symbols, used throughout the subsequent text and figures:

$a$      the allometric coefficient of change of any process with body mass

$\alpha$      a constant defining scale in the empirical relation of rate of metabolism and body mass

$A_\nu$      normalized biomass spectral density function with body mass in Model I

$B_\nu$      normalized biomass spectral density function with body mass in Model II

$\beta_\nu$      unnormalized biomass spectral density function

$B$      biomass density

$B/P$      the average turnover time (inverse of specific production)

$B_s$      average body mass density for species $s$

$C$      prey biomass captured and consumed by predators

$c$      coefficient of curvature in the quadratic part of the solution of Model II

$\gamma$      coefficient of the rate of metabolism in relation to body mass

$D$      deaths due to nonpredation causes

$E$      rate of food-energy elimination (indigestible consumption)

| | |
|---|---|
| $F$ | rate of predation |
| $f$ | the number of units of fishing or predator effort |
| $G$ | net rate of population growth |
| $H$ | periodic terms in the solution of the biomass function for Model I |
| $J$ | periodic terms in the solution of the biomass function for Model II |
| $i, j$ | subscripts specifying a body-mass group as prey or predator, respectively |
| $K$ | ($= m \times p$) the efficiency of converting prey to predator substance |
| $M$ | rate of natural, nonpredatory mortality (called natural mortality in fisheries) |
| m | metabolic efficiency |
| n | trophic position |
| $\nu$ | any body-size interval of the body-mass distribution |
| p | digestive efficiency |
| $P$ | rate of production of the biomass density |
| $P/B$ | the specific rate of production |
| $P/B_s$ | specific rate of production for species $s$ |
| $\pi_\nu$ | unnormalized production density spectral function |
| $q$ | "catchability" coefficient (probability of capture by one unit of effort) |
| $r$ | individual ration |
| $R$ | ratio of predator to prey body mass |
| $S$ | assimilated energy |
| $T$ | rate of respiration of the biomass density |
| $w$ | body size (measured in mass units) |
| $dw_\nu$ | a small body-size interval ($\nu$) specifying a given unit of biomass density |

PREFACE

࿄

Developments in ecology over the past 25 years have been an unexpected source of wonder and satisfaction to us in our synthesis of knowledge of the processes underlying production in aquatic biological systems. What has become clear through this undertaking is how vigorously the appearance of new conceptions continues into the present. As a result, we have found both old and new theories to be relevant to understanding in the field of ecology: in fact, some of the tools essential to the development we undertake here are so new that we have come to believe that this synthesis could not have occurred much before now. We have, therefore, dedicated our work to two of our teachers of ecology who were among the first to demonstrate to us the importance of trying to remain open to the new perspectives continually being revealed.

Dr. A. G. Huntsman, whose career as biologist and oceanographer spanned nearly half a century and included the founding and direction of several laboratories for the former Fisheries Research Board of Canada and a teaching career at the University of Toronto, struggled for many years to find a formulation for the characteristics of organisms that could include the environmental context of their behavioral life histories and their physiologies in one measurable whole. Early in his career (Huntsman 1918), he outlined a numerical model of fish population age structure as a function of the rates of annual recruitment and mortality, intending it as a means of illustrating the simple logic that dictates changes in abundance in relation to exploitation. His system of calculation was parallel to that published in the same year in Russian by Baranov (1918, although not available in English translation until the 1930s). However, Huntsman did not continue his analysis in the direction of Baranov's important simplifying idea that incorporation of a growth curve could convert his steady-state numerical model into a theoretical predictor of relative biomass changes. Instead, as is shown in two of his later, more philosophical papers (Huntsman 1948, 1962), he was concerned by the fact that, because we cannot understand and predict the actions that result in changes in the movements and distributions of organisms, we cannot usefully predict their abundance changes in the regions that interest us. He believed that to serve our practical purposes we needed a means of observation that could objectively associate properties of organisms with the changes they continually encounter in their environments. To emphasize the importance of these ideas, he coined the two faintly ugly terms *biapocrisis* and *ectology,* which he defined, respectively, with the elegant simplicity of which he was capable, as "the response of the organism as a whole to what it faces where it lives," and its converse, "how the environment determines what happens to the organism." The ecology of that period did not find his formulations or his terminology useful as a basis for prediction. However, Huntsman's perception that emerging questions of ecology required for their practical solution a new, larger, and more inclusive scaling of the units making up the ecological universe was ahead of his time. It is a perception to which we have returned only as we have begun to appreciate the deficiencies, related to the actual distributions of potentially interactive elements, that have shown up in a variety of unsuccessful attempts to apply simplifying, homogeneous numerical models to the prediction of aquatic production.

Dr. F. E. J. Fry, professor at the University of Toronto and director of the Ontario Fisheries Research Laboratory, was a student and independent appreciator of Huntsman's reasons for groping toward new breadths of conceptualization of aquatic biological systems. He drew his students' attention to the necessity of differentiating between the objectives for an autecology and a synecology of ecological systems (Fry 1947). He saw that these two approaches required adjusting the scales of observation and analyses to the purposes of any given investigation. In his synecological approach to population field work, he invented the "virtual population analysis" (Fry 1949), which became the basis in fisheries science for calculating year-class abundance from the sequence of annual samples obtained by the sampling of commercial fisheries catches. In the laboratory, as we indicate in more detail in chapter 6, Fry took a series of new and imaginative autecological steps toward a precise and quantitative "holistic" perception of organisms in relation to environmental "factors" as a means of understanding changes in distribution and abundance in present and past environments. His view of the essence of ecology embodied the concept that a system is energy that is "contained or channeled through form [where] . . . the fundamental definition of form is a pattern of forces" (Kerr 1990). In these complementary ways, these two thinkers sought to link the organism with its environment in a manner that combined the growing use of mathematical formulations with the burgeoning information on animal behavior and physiology, to which they made notable contributions.

While we trace our initial realizations of the possibilities of this inquiry into aquatic production to the influence of Huntsman and Fry, there are a number of other colleagues whose influence on our work has been more specific, and whose contributions we want to recognize here. We recollect with particular delight the debates of the late 1960s and early 1970s among the staff of the Fisheries Research Board's newly formed Marine Ecology Laboratory at the Bedford Institute of Oceanography. All were intrigued by the possible general nature of production systems implied by Ray Sheldon's persistent compilations of the size compositions of the plankton in his primitive Coulter Counter results. Sheldon's cooperation with Bill Sutcliffe played a major role in applying the results to the larger organisms in the natural environment. There were very diverse evaluations of the results in those early days, but the ferment of ideas gave rise within the same laboratory to two main lines of investigation: that initiated by Trevor Platt and Ken Denman, and the work of Peter Schwinghamer and Gary Sprules in connection with the interests of the present authors. At various times these individuals, and others, including Paul Boudreau, Martial Thiebaux, Jyri Paloheimo, Bill Silvert, and a number of interested students, Martha Krohn in particular, have contributed to the development of our ideas in constructive ways that have stimulated our efforts and promoted reformulations. Dale Webber, as always, provided much useful support. Especially important contributions in terms of ideas, methodology, data, and additional analyses have been made more recently by Daniel Duplisea, now in the Fisheries Laboratory at Lowestoft, who has also been kind enough to read and criticize early versions of much of the present text.

We want to take this opportunity to especially acknowledge the support offered our earlier efforts in discussions of the nature of dynamic systems in biology with the late Robert Rosen. He read and discussed with us some of our earlier explorations of the application of his concepts of systems to fisheries ecosystems and actively encouraged us to pursue the ideas further. We regret that we have been unable to share with him our satisfaction with some of the results.

The mythology of science attributes advances to the efforts of specific individuals. However, the record of contributions to the work that we present here convinces us that science really depends on the combined efforts of many individuals. We are deeply appreciative of the unselfish offers of advice and support from many friends and associates, representing a diverse and broad spectrum of experiences. They have made our synthesis more stimulating and rewarding than could ever have been conceived without them.

S. R. Kerr
Cox Fisheries Scientist-in-Residence
Biology Department
Dalhousie University
Halifax, N.S. Canada B3H 4J1
srkerr@is.dal.ca

L. M. Dickie
Scientist Emeritus
Department of Fisheries and Oceans
Bedford Institute of Oceanography
Dartmouth, N.S. Canada B2Y 4A2
dickiel@mar.dfo-mpo.gc.ca

# THE BIOMASS SPECTRUM

# CHAPTER I

ॐ

# Choosing and Organizing Observations and Experiences

In this book we propose a development of ecological theory that we believe can guide a more effective remedial treatment of the drastic perturbations in natural production systems caused by heavy exploitation and environmental disturbances from society. This development is based on empirical evidence of the stability, in different times and places in the world oceans, of what we call the biomass body-size spectrum. Its characteristic shape and structure enable us to perceive an underlying ecological energetics, which we hypothesize derives primarily from predator-prey actions of energy acquisition and transmission in food chains. The main advantage of invoking the predator-prey model is that it permits us to develop new measures of parameters of energy exchange that are directly affected by exploitation and environmental disturbance. It thus becomes a potentially important addition to the tool kit of those charged with environmental and natural resource protection.

Because of the ordered structures that emerge from the natural dominant hierarchical arrangement of body sizes in aquatic food chains, we are currently best able to perceive in them that changes in biomass spectral structuring are responses to particular causes of perturbation. Therefore, our main attention in this book is directed to aquatic systems. Effects in terrestrial systems are more difficult to interpret. However, where we have the evidence, we discuss applying the same principles to them. We believe that a comparable assembly of biological data for all natural production systems will eventually reveal the generalizability of the approach we take and that it will lead to a more complete understanding of the limits of all natural production systems to adapt to human intervention. Such understanding is prerequisite to developing the restorative measures currently needed. In retrospect, the biomass spectrum is the almost inevitable outgrowth of the present-day recognition of a need for more broadly responsible human actions to meet the commonly called-for objective of sustainable development of the world's "renewable" resources.

In their recent book *Hierarchy Theory,* Ahl and Allen (1996) produce an aphorism that aptly describes what we believe to be the crux of development of general scientific theory as surely as it is the crux of development of the ecological theory we undertake here. As they put it, "Observation is structured experience" (51). This perception of the processes at the very base of scientific discovery embodies the necessarily radical new approach of all late-twentieth-century science to identifying and interpreting the factors of observation that can enable us to determine, with acceptable accuracy, what the future holds. Especially over the past half century, as burgeoning technique has opened new experience, observations have multiplied at bewildering speed with bewildering complexity. In this process, the earlier concept that observed phenomena can be directly related to simple causes has long given way to the realization that the world of scientific interpretation depends as much on the concepts and methodologies with which the observer approaches the object as it does on the perceptions that are called observations. As has been pointed out in many recent popularizations of science, in particular in physics (e.g., Zukav 1980), interactions of conception and perception are capable of comprehending a virtually endless variety that denies the once sought-after uniqueness of a simple relationship between the observation of results and the identification of causes in an objective, external world.

Consequently, a principal feature of our discussion of ecosystem properties in this book is the formulation of well-defined observations on appropriate scales of aggregation within a hierarchical causal structure suited to the purposes of the investigation. In retrospect, our central idea of plotting the densities of the biomasses of all organisms in a community by body-size classes in the biomass spectrum has come to seem an obvious device for generalizing about the nature of ecosystems. It is the hierarchical pattern of interaction revealed by this method of compiling observations that has supported development of the underlying theory of predator-prey energy transfer and has led to a new understanding of ecosystem dynamics and of the controls and limits on its responses to perturbation. In fact, however, this method of compilation and analysis has been slow to arise and is only one alternative, and a late one at that, among the analytical approaches that have been considered suited to the development of ecological knowledge. As we point out in what follows, the biomass spectrum approach is an alternative that is still so new as to require considerable further testing in both space and time, yet the conclusions it has already afforded cannot be ignored.

We wish to show the newly perceived potential of the biomass spectrum methodology for achieving a broad and comprehensive understanding of the dynamic processes underlying production and its exploitation, especially in aquatic ecosystems. We demonstrate the explanatory powers of the methods both in understanding the internal dynamics of organism interaction in aquatic ecosystems and in predicting how the whole aquatic community interacts with measurable features of its external environment. In this we are particularly concerned with examples of the effects of heavy fishing on natural communities of organisms in the sea and in freshwater ecosystems.

While the methodology offers new methods of compiling data, most of the basic data required already exist in fisheries laboratories in various parts of the world. That is, the main requirement in many cases is a simple reformulation of existing data series. This reformulation alone should be sufficient to establish the possibilities for new measures of community perturbations and to find out how they can be used to more precisely determine the impact of various forces. The conclusions that we reach from our own reformulations offer hope that a broader and more effective system of remedial action can be undertaken in many natural production systems. However, we hasten to add that in fisheries it appears that a satisfactorily efficient remedial system will require a major restructuring of the administrative systems charged with regulation. Analyses of the biomass spectrum afford perceptions of factors determining ecosystem production that call for new concepts to be applied to the structure of regulation systems and new objectives to be assigned to the elements in that structure. In the final chapters of this book we discuss the particular application of these perceptions to fisheries-related control systems.

## General Methods of Ecological Analysis

An earlier form of ecological analysis, based on enumerating the quantities of various species and types of organisms over time, still persists as the methodology of choice for much current study of aggregations of organisms, and most systems of ecosystem control are based on conclusions drawn from it. This direction in both research history and in attempts to regulate population production depended on the early mathematical formulation of logistic population growth by Lotka (1956) and the subsequent development of numerical predator-prey interaction models by Lotka and Volterra. Much of this work originated from analyses of the spread of malaria among workers during construction of the Panama Canal. In aquatic systems some of the management potential inherent in this approach was directly adopted in Schaefer's (1967) model of fisheries and in his application to analyses of data on the declining catch of Pacific tuna fisheries in the face of increasing fishing effort following World War II. Even earlier, in fisheries, however, was the simple and clear formulation of Baranov's (1918) numerical model of the stable age distribution of successive broods of newborn fishes subjected to various rates of fishing. This was initially presented by him as a way to bring order to the confused discussions of "overfishing" that were already being argued in relation to apparently declining fisheries of the North Sea. Elaboration of his basic methods, utilizing various theories of the growth rates of the individual species involved, was undertaken by Ricker (1940, 1975) and Beverton and Holt (1957), and the same numerical approach was utilized in Fry's (1949) virtual population analysis. Elaborations of these numerical methods are currently widely accepted as the most reliable basis for analysis and management of changes in species yields in the great fisheries of the world.

In developing numerical methods, fisheries population biologists have implicitly accepted the proposition that populations are made up of species represented by variable numbers of individuals. In models of these individual species, the numbers usually need to be classed secondarily according to their different characteristics of size or age. Enumeration by species seemed particularly obvious as a scientific approach to generalizations about production when the organisms being considered were mosquito-borne parasites or bacteria, which could be regarded for this purpose as identical particles living in a homogeneous and stable environment. But even for multiaged populations such as those of fishes, as pointed out by Baranov, the only real difficulty in calculating production this way seemed to be uncertainty about how to incorporate a suitable growth curve among the numbers. Subsequent analyses in fisheries-related systems recognized the need to also include control of reproduction (Ricker 1954); but the basic underlying principles of population change, elaborated in matrices of increasingly sophisticated design, are still those implied in Lotka's original numerical explorations of the struggle for life among the sums of homogeneous particles in a uniform environment.

The ideas and observations propounded by the school of E. P. Odum (Odum 1969) can be credited with the most effective perception of the aptness and viability of a complementary, energy-based system of analysis. It is the line we have followed here. In ecological systems of different types, Odum and his associates perceived that measurements related to the energy flow through ecosystems can be directly utilized to understand and predict the dynamics and organization of the processes that control ecosystem production, and that the conclusions may be different than those afforded by either numerical or biomass calculations. These generalizations were amply illustrated in early studies of the salt-marsh flats of Sapelo Island, Georgia (Odum and Smalley 1959), on how the variances of biomass and numbers of organisms showed different aspects of the well-known patterns of seasonal change. When, however, these results were translated into energy terms, they could be seen to reflect the well-known metabolism-body-size rules that govern individual organism respiration and production in the communities. The more orderly results that came from this methodology led directly to an understanding that measures of organism body sizes and densities together provide an index of productive capacity related to the effectiveness of the population's utilization of their available food-energy supplies. That is, calculations using the energy methodology introduced the possibility of using the total energy balance within the system to interpret responses of its dynamics to various factors. The importance of the concept of an energy balance could not be directly deduced from the numerical methodologies.

The additional recognition that there are complementary effects of body mass and respiration with individual body size, and that these relations apply broadly over species, led to early explorations of the general relations between the energetics of production and the aggregations of body sizes of organisms in communities. Aspects of population growth and reproduction of many species in relation to body size were compiled early by Smith (1976), suggesting relationships that have been amply confirmed in allometric studies in more recent years (e.g., Peters 1983). Even these early results introduced the further important possibility that understanding and predicting production may be feasible at the level of analysis of the broader properties of the affected communities and need to be characterized in relation to human interventions, without first separating the data according to species.

It is in this current of ecosystem studies that the usefulness of the biomass spectrum of body sizes in providing measures of ecosystem productive capacity has emerged. As we point out in what follows, the consequences of this line of inquiry, initiated in oceanography by the work of Sheldon and Parsons (1967) and by Sheldon and associates (1972, 1973), have led to an appreciation of the energy conservation balance that underlies ecosystem change. We believe that the biomass spectrum thereby shows itself as a powerful adjunct to the tool kit of all ecologists, even though, up to now, its application to terrestrial ecology has been sketched only in the barest of terms. In aquatic systems, perceptions of its application to population control are more broadly recognized, but they have yet to be taken seriously in the practical management of human exploitation.

In the course of this exploration we have also come to better appreciate the important role of body size itself as a criterion for generalizing on biological dynamics. An impressive array of books, based on a large journal literature, has directed attention to body size as an index of various performance attributes in relation to ecosystem properties. While some have emphasized its simple correlation with various physiological and behavioral attributes, others have embraced a broader functional analysis of body size and metabolism in relation to performance characteristics of both individual organisms and population densities. They have thus demonstrated its importance at various levels of biological organization, including both physiological and ecological processes. This background has enabled us to explore the usefulness of indices of body size as an alternate means of classification, especially appropriate to a hierarchical structuring of the ecological analysis of production systems.

This is not to say that we do not find a need to make use of taxonomy in relation to community production. However, in what follows here, much of our use of this oldest of the branches of biological study is of a quite general nature. Readers with a more exacting taxonomic bent will, we trust, excuse our incessant turning to "fishes" or "phytoplankton" as though these constituted familiar and well-defined group-ings of aquatic organisms. Our trust is based on the fact, which will be clear enough, that this usage is often sufficiently precise for our needs. In general, we require taxonomy more for referring to groupings of or-ganisms in relation to their trophic positionings in the community than for a precise defining of attributes of organismic vital statistics in pro-duction studies.

Given recent trends, we feel we must attach an important qualifica-tion to these statements. Nothing in our analysis should in any sense be construed as an attempt to *replace* conventional research in taxonomic classification. For reasons that escape us, funding for taxonomic research has been selectively decreasing at the very time when events related to biological sensitivities of different types of organisms to environmental change show that we need it most. For this reason, we explicitly disavow any possible interpretation of what follows as an excuse for adopting the position that taxonomic research is an obsolete science that no longer merits support. Our view is quite the contrary. We regard the size-dependent approach developed here as a necessary device for studying the generality of the organization of organisms into an ecosystem. How-ever, real understanding of function in the community also requires an ability to refer to additional biological criteria of organism type as the need arises to distinguish between particular species, or races, within a system. We discuss this in some detail in chapter 8. Thus, a method of analysis of production that employs body size as a primary classificatory criterion is a complement to conventional typological analysis, not its replacement.

## Organization of This Book

In approaching our analyses, we need to make explicit our preconcep-tion that estimating production of a whole ecosystem by summing the functioning of its currently dominant taxonomic elements may not ad-equately reflect regularities in the total system energy flow. In fact, the wide experience of unexpected changes in the representation of species in ecosystem productivities has been an important influence leading us to this study. Nevertheless, in chapters 2 and 3, statistical description has enabled us to characterize the biomass spectrum of a number of aquatic ecosystems on the basis of the immediate body-size composition of constituents and to discern a compelling similarity of pattern among systems: namely, an unexpected persistence of hierarchical patterns in the body-size structure of biomass density at certain levels of spatial and temporal aggregation. This evidence implies both a special need and a special opportunity to make the causes of such structure explicit in terms of mechanism.

Our view is that observation of pattern in the attributes of a complex system at a given level of aggregation of observations implies that regulation of the underlying processes takes place at that same level of scaling. Evidence that different patterns emerge at different levels of aggregation leads us to search for a hierarchical arrangement of the forces that might affect ecosystem elements at different, possibly quite independent, levels of influence. Considering the effects at different levels of the hierarchy, and comparing conclusions from these, evokes interest in more than one set of explanations. The results inevitably involve causes that are sometimes complementary but often contradictory in their simple effects. Their combined results are necessary to understand the principal components of the underlying dynamics. These ideas are explicitly developed, beginning in chapter 3 but continuing into chapters 4 and 5.

In chapter 4 we are most concerned to lay out a mathematical formulation of a common predator-prey dynamics that we infer may underlie the observed, size-dependent properties of biomass spectra at the different levels of the hierarchy. This formulation allows us to compare the observed structural characteristics with those deduced from the mathematical construction, and leads to a more extensive study in chapters 5 and 6 of the ramifications of these same mechanisms.

In chapters 5 and 6 we also put forward a more extensive size-dependent approach to the mathematical description of production systems in terms of a hierarchical scaling of the ecological and physiological mechanisms identified in chapter 3. We base this hierarchical arrangement on what appears to be a strong principle of similarity underlying common interactions among the predator-prey body-size elements at successive steps in the laddering of the predator-prey elements that are the carriers of the energy flow through the ecosystem. This arrangement allows explicit representation on different but identifiable scales of some of the important mechanisms responsible for the observed patterns of structure operating in biological systems. Identification and mathematical specification, in this case at the levels of both individual and population interaction, without primary specification by species, leads to a formal inference of the existence of a number of the ecosystem phenomena that appear in the more descriptive chapters 2 and 3.

Readers unfamiliar with the journal literature concerning the biomass spectrum may need to read chapters 2 and 3 with some care if they are to be convinced that we are addressing real phenomena. Similarly, readers who have not particularly followed the development of biomass spectrum theory over the past three decades may want to peruse chapters 4 and 5 to inform themselves on what has been achieved in relating observation to a consistent body of theory. These chapters are primarily reviews of the existing journal literature, fortified in chapter 6 by a more detailed consideration of the physiological mechanisms on which spectral phenomena are based.

Those familiar with the biomass spectrum literature might prefer to start with chapter 7, where we explicitly begin considering how observation and theory may be related to the elements of system interaction and how external influences affect internal dynamics. At this stage, if we have persuaded the reader that the size-dependent phenomena we review are real and ubiquitous, and that the observations flow from a sound theoretical expectation, the reader may begin to share a sense of the beauty and elegance of how natural ecosystems behave in the size-dependent domain.

Of course, our interest is more than aesthetic: we believe that the interactions identified have important practical consequences for changing the view of what constitutes effective management of aquatic production systems. To this end, the final chapters of this book address the task of providing an ecological context for the application of biomass spectrum theory to the pressing problems of ecosystem management, especially in fisheries. The content of these remaining chapters is more substantially dedicated to synthesizing recent knowledge of system behavior in relation to biomass spectral views of aquatic community changes than to simple collation and summary. Of course, we need to continue to draw heavily on the extant journal literature. It follows that we hope that our most interested readers, even those familiar with the biomass spectrum literature, will read and judge this material in its entirety.

Chapter 7 initiates a system context for the ecology of the biomass spectrum. For this purpose, we return to some particular ideas about system organization that fall under the rubric of hierarchy theory. Our primary tool is a study of what might be called "system constraints," which are the consequences of the asymmetric interaction of slow and fast variables between hierarchical levels of complex systems. The essentially metaphorical approach offered here may sharply contrast with the more closely reasoned mathematical development in preceding chapters. It is, however, based on the rigorous mathematical development outlined by Rosen (1991) in the first book in this series and subsequently extended in his recent book of essays (Rosen 1999). Our major purpose is to focus on the ecological consequences of the mathematical developments of chapter 4 when they are placed in the context of general system behaviors, and which by themselves would have constituted little more than an interesting intellectual exercise. But in considering this general problem we necessarily turn to some of the seminal ideas of contemporary systems theory. These are often considered under such categories as "self-organization," "dynamical systems theory," "nonlinear dynamics," "complexity theory," "dissipative nonequilibrium systems," and other variants on the same theme. What these approaches share in common is their recognition of the remarkable properties that *organization,* as distinct from structure or function, confers on biological systems. These ideas are so new, so untested, but so fascinating that we necessarily tread this terrain carefully. With attention to the possible consequences, we focus particularly on what Rosen called the "semantics" of biological systems.

Biological or ecological processes are amenable to various analytical procedures. In chapters 8–11 we develop the view that size-spectrum theory has reached the point where it has much to offer in helping us perceive what is needed for management of fisheries ecosystems. We therefore address some of the obvious potentials for management offered by the developing methodology of size spectra. To accomplish this, we examine alternative ideas about the organization of aquatic production systems.

For some appreciable time, phenomena related to the orderly, size-dependent decrease of specific metabolic rate with increasing body size have captured the attention of biomass spectrum theorists and empiricists alike. However, attention to these phenomena neglected what we believe to be the important secondary structures (domes) in the spectra. The placing of these latter phenomena in context, in the light of hierarchy theory, enables us, in the final chapters, to perceive the consequence of the basic organization of aquatic ecosystems in relation to the various types of causes that appear at different scales in the system.

Our ideas appear congruent with those such as Johnson (1994), who has described the properties of depauperate lakes as "reference systems." We differ only in terms of the system complexity that we find can be comprehended in recent observations of aquatic ecosystems in the world. Given the sensitivity with which ecosystem dynamics are revealed by the parameters of the biomass spectrum and the similarities between observed systems, we are able to examine the effects of various internal and external factors on productivity. In particular, it appears that our approach allows a new and more operational view of the effects of fishing mortality in shaping the response of multispecies fisheries to the unprecedented rates of fishing that have developed in recent years.

The results of this exercise have significant implications for the management of multispecies fisheries. We conclude our account by spelling out in detail what must be taken into account for an effective implementation of an ecosystem approach to fisheries management: things essentially lacking in current approaches to the problems of management.

## Other Considerations

To a certain extent, the fact that we spend much of our attention in this book on fishery production systems as a special aspect of aquatic biological systems comes from our backgrounds: this is the literature we know best. As stated earlier, however, we believe that our statements about aquatic production systems apply with equal force to their terrestrial counterparts, and so we refer to the potential linkage in several places. However, the nature of the distribution of body size among the primary producers in the terrestrial community places considerable difficulties of detail on those who would extend these ideas to terrestrial systems (Griffiths 1998). With some reluctance, we have concluded that because many of these details seem not yet to be well known, it is necessary to leave it to others to flesh them out. We can see no difficulties in principle in such extension, but it may take considerable time to develop better means of survey. It is our good fortune that the aqueous environment has readily afforded a reliable methodology for our studies.

In similar vein, we have dealt largely with fishery exploitation as the primary anthropocentric source of stress to aquatic production systems. This is a device we chose for convenience and clarity, but also because fisheries worldwide appear to have been subjected to severe stresses that have not been confidently relieved by current "conservation" measures. Because of this they deserve early and special attention. However, it is clear that additional sources of human-induced stress are rapidly growing in importance in natural production systems. Some of them—for example, the effects of eutrophication, species introductions, contaminants, climate modifications, and a variety of associated stressors—could have been dealt with in much more detail. We have every expectation that the impacts of all these variants of human interventions on natural production have much the same patterns of effects as are perceptible in the fisheries. Had we chosen to deal with the full slate of human interventions in natural production, even within aquatic production systems, this book would have been much longer than it is, without essentially adding to its insights into the types of discernible mechanisms and results. We believe that fishery exploitation is a useful general model through which to detect and measure the nature of most if not all of the perturbations humans have imposed on natural production systems.

We should also note that an "ecosystem" approach to management of natural systems has been increasingly called for by those wishing to emphasize their discontent with the present state of management of the world's fisheries. This call has been interpreted in some quarters in a positive operational sense (e.g., by the Great Lakes Fishery Commission between Canada and the United States) as a call to truly reexamine basic approaches, including, in particular, single-species management and its associated problems. Unfortunately, however, some very large and important national as well as international management agencies seem to have felt free to interpret these calls as so much rhetoric, to be disregarded by those intent on continuing business as usual, who make only slight changes in the intensity of selected sampling methodologies, or simply provide old techniques and methods of analysis with new or different names. We hope the ideas set out here can clarify and define methods sufficiently to help redress such irresponsible promulgation of the comfortable illusion that we have seen it all before and that, given time and persistence, all will be made well again through existing institutions. Such attitudes on the part of administrators of some of the major natural resources of the world must come to an early end.

We are ready to argue that the synthesis offered here is indeed a new and viable "ecosystem approach" to the analysis and management of aquatic production systems: the results call for substantial changes in important aspects of present management principles and procedures. We wish to emphasize, however, that this approach is firmly rooted in known and measurable ecological processes, and its expression in the form of the biomass spectrum appears to be the inevitable consequence of the ways that natural systems interact. Our methods and their application may not prove to have all the salvaging powers that we wish for them, but they will, at the very least, provide an alternative springboard for developing and applying the more sensitive and broadly based methodologies urgently needed.

∾

# CHAPTER 2

# The Empirical Evidence

The earliest presentations of the biomass spectrum were made by Sheldon and associates (1972, 1973). They were based on plankton samples taken in the pelagic zone of marine environments and compiled by methods devised earlier by Sheldon and Parsons (1967), using a modified version of the Coulter Counter for data collection. They were concerned with interpreting an overall regularity that they perceived in the pattern of biomass concentration by body size. This was based on plots of their own phytoplankton samples as the logarithm of the concentration of biomass in successive logarithmically ordered body-size classes, augmented by data from the literature to include average concentrations of larger body sizes of organisms. From this series they concluded that the whole spectrum was surprisingly "flat."

It was not long before samples from other areas and environments were studied in similar fashion by other investigators. In the course of these studies, interest in the discernible patterns seemed to diverge along two main lines: one, like that of Sheldon and associates, was concerned with the simplicity of an overall trend; another was more concerned with a complex, though regular, internal structure underlying the simple trend. In retrospect these two approaches seem to represent different attitudes to scientific analysis of empirical data, and for this reason we review them separately here. We call the two approaches the "integral" and the "analytic."

The integral studies attempted to estimate the slope and intercept parameters for a single spectral line constructed over the whole range of organism body sizes that could be sampled from a given environment. In general, investigators following this approach concluded that the spectrum for a particular environment could be represented by a straight line of low, negative slope. By contrast, as some investigators noted the repeated, often distinctive domelike structures in the whole spectrum, they concluded that the integral overall linearity may be a simplistic result of averaging over domes at smaller scales—within segments of the entire body-size range. That is, the integral simplicity arose from drawing lines through averages of significant underlying structure and dynamics within the whole spectral range. Emphasis on the analytic perception of a substructure implied a need to calculate parameters for a more complex set of forces and scales of action than was called for by the integral approach. The implied question was whether causes at these two scales of compilation were related to one another, or represented independent dynamics within the whole system.

In what follows in this chapter we direct attention to the integral spectrum, where emphasis will be on estimating and interpreting the parameters of an overall fitting of the data. Reviews of the contrasting analytic approach are begun in chapter 3, with a consideration of the details of the internal structural pattern and its possible causes. Inevitably, organizing the background data and discussion into two different chapters involves a somewhat artificial separation of the lines of argument developed in the literature; in fact we have concluded that an intertwining of elements at both the overall and the detailed levels is necessary to the comprehensive interpretation of the dynamics we undertake here. However, differences in the underlying rationale for constructing and interpreting the two views of the spectrum are clearer if we start with this two-pronged approach.

Bear in mind that over the same period that the spectrum has been developed, there has been a changing awareness of the significance of body size as an index of the dynamics of biological systems generally. We therefore begin this chapter by considering some of the main themes of investigation of body-size indices and their implications for classification of the spectral data by body size. We also undertake a short digression into the limitations of perception that are inherent in some of the technologies of data collection and analyses that have led toward recognition of the spectrum. In analyzing aquatic systems, we are necessarily concerned with solutions to instrumentation problems on which the data depend. Following this introduction, we begin reviewing the results of the data compilations themselves and their interpretations.

## Body Size as an Index of Ecological Interaction

The line of inquiry linking the productivity of aggregations of organisms to various attributes of body size has its roots in early ecological concerns about how habitat is shared among the many types of organisms that inhabit it. Models of the kinds and numbers of organisms by Hutchinson and MacArthur (1959), and the related "broken-stick" models of MacArthur (1957, 1960), designed to explore the basis for the observed relative abundances of species, clearly showed that within a given habitat the body-size distributions within groups of similar species are a function of resource allocation among them. Similarly, Schoener (1974) recognized that regularities in the density of predators reflected in the territory sizes of birds derives from relations between body size, food density, and the related efficiencies of their various feeding behaviors and strategies. Thus, body size, through its relationship to trophic interaction, emerged as an important index of the organization of ecosystems. These perceptions of a relation between organization in communities and the densities and functions of individual organisms were important landmarks in exploring the basis for the ecological distributions of organisms and the productivities of various physiological, functional, or behavioral types.

The scope of subsequent analyses enlarged to show that body size may reflect a variety of causes. For example, Sprules (1980, 1984) showed that classification of zooplankton communities according to body-size distributions provided a more powerful predictor of production differences between freshwater communities from different zoogeographical and lake-morphometry classes than did classification by conventional species types. Subsequent extensive compendia of correlational associations of body size with various physiological and behavioral indices were made by Peters (1983), Calder (1984), and Schmidt-Nielsen (1984). They used the concept of body-size allometries to index a wide range of both individual organism and population attributes. These allometries gave abundant evidence of an underlying regularity in the ways functional relations are distributed among the body-size components of multispecies populations.

Allometric relations had originally been seen as means of appreciating the commonality of life functions that are expressed in diverse body forms (Thompson 1917), or of comprehending the seemingly diverse growth rates and physiologies of various body parts into functioning whole organisms (Huxley 1932). The compendia of Peters and Calder, in particular, verified that physiological and behavioral functions—ranging from basic metabolic rate and locomotion of individuals to reproductive performance and life history traits of populations—all have common features or characterizable differences in relation to body size. Recent advances in practical and theoretical allometric studies have now demonstrated that the numerical values of allometric exponents offer new insights into the basic biological mechanisms involved (West, Brown, and Enquist 1997; Enquist, Brown, and West 1998). These have consequences for the interpretation of biomass spectral parameters that we will discuss later in this book.

From these allometric studies we may infer at this point two facts of particular importance. In the first place, the allometries that characterize the biomass of broad species groups may have different values than those found within individual species. From this it appears that a body-size index provides a classification of functioning of aggregations of organisms on different levels of integration than those provided by conventional taxonomic classifications. Since our purpose in this book harks back to the questions of Schoener (1974) and Sprules (1980) about how the sustained productivity of different ecosystems is realized through the densities of organisms that are aggregated in them, an index that reflects the rates and efficiencies of energy flow at various levels of aggregation is an obvious means of studying the dynamics of communities.

In the second place, as shown by both Sprules (1980) and Peters (1983), because different levels of aggregation of observations of body size reflect different aspects of organization of the energy flow through communities, a specifically ecological application is embodied in a classification system that employs body size as a primary criterion of classification. That is, if we are to understand ecological attributes such as biomass, production, and yield from natural ecosystems, we must know the underlying dynamic factors on a functional rather than a purely taxonomic basis. A body-size classification is clearly a more appropriate taxonomy for production studies than a species classification (see also chapter 8). We adopt it as a criterion of organism type specifically suited to the study of energy flows and productivity at the ecological level of biological generality.

## A Hierarchical Arrangement of Ecosystem Biomass Elements

It is appropriate to take account of the usefulness of different scales of classification of the biomass elements for discerning the factors controlling ecosystem function. Units of observation always have to be considered in various degrees of aggregation or organization in relation to the dominant time and space scales that may be affecting them. Some index of the total living biomass, such as its biomass density (mass or energy per square meter of environment), is logically the most general level of observation at which we could expect to find regularity in rates of production. However, in aquatic studies this is more generally regarded as informative when applied to the study of parts of ecosystems, such as where the investigator determines the concentration of total phytoplankton or zooplankton in the pelagic layers of an aquatic environment; although, even at this general level, the discernment of biologically meaningful pattern in the production and distribution of biomass appears to be enhanced if biomass is subdivided according to species and/or body-size categories during distinct time periods (Cottingham 1999). It has often been pointed out (Platt 1985; Dickie, Kerr, and Boudreau 1987) that this is directly related to the strong body-size dependence of metabolic rate. This broadest level of determination of biomass is therefore usually subdivided into a second level of functional types defined by body-size categories, which we may regard as occasion demands as a second hierarchical level of observation of the total system.

At these more detailed, intermediate subgroup levels, we may expect to find regularities in biomass concentration or densities assignable to environmental types. For more detailed ecological analyses, designations of life history stages of organisms as predator or prey types may be equally valuable, since ecological studies are inevitably concerned with the organization of community and population elements in relation to nutrients and energy flows. In these cases, regularities appear more clearly when the group designations are defined according to body-size ranges in relation to their functions in the ecosystem.

At the finest hierarchical level of ecological observation are the individual organisms with their individual physiologies and behaviors. Body size is clearly an important attribute useful for classification of the biomass at this level as well; and while single individuals can also be assigned to species, an understanding of their functional role in the community equally often requires classification by other attributes. Sometimes finer divisions of biomasses, by age groups or behavioral types, need to be investigated in relation to metabolic efficiency or susceptibility to noxious chemicals. Coarser definitions, such as assignment to particular predator or prey types, are of advantage in investigating basic production. It is thus evident that in ecology we need to invoke various degrees of aggregation of organism attributes in relation to nutrient or energy flow. Among them, body size serves as a useful criterion of classification at several levels of the possible hierarchical organization of the observations of biomass elements.

From the point of view of ecological analysis, purely taxonomic designations of groups of organisms are most often relevant as convenient identifiers of particular functional subgroups, of which more than one may constitute a particular ecosystem component. It should thus be clear that compilation of the body-size spectrum of the biomass of all organisms in a given environment becomes advantageous particularly when we begin to seek information on what controls the total energy flows and how the overall production result depends on the various partitionings of the total biomass into component parts. Physiological mechanisms regulating energy flow are developed and appear to reach certain efficiencies through natural selection acting on individuals of species in particular relation to others of the same species. But these mechanisms operate ecologically in the environment in relation to the energy supply that comes, first, from the incidence of sunlight on the total surface area of the environment. This energy supply is then distributed among the various organism types in the organized ecosystem according to their trophic functions in that system. For this reason in particular we also specify biomass density at the various degrees of aggregation in terms of its concentration, calculated as a density per square meter of the surface of the environment in which it is sampled.

It follows that a body-size classification of biomass that permits flexible aggregation of biomass observations is suited to the search for evidence of effects that may arise at various levels in systems. We need to know at what spatial and temporal scales observations of the effects of interactions at the various scales of aggregation of organisms can be turned to our analytical advantage.

## The Dependence on Technology

Pursuit of the patterns in which biomass and production phenomena are manifested requires attention to the nature of the evidence as well as its time and space scales. Since general evidence for body-size dependence of production is ubiquitous and has existed for a considerable time (Engelmann 1966), the need here is to review current perceptions in relation to the technology used to reveal spectral characteristics. Particular methodologies impose their own limitations on the generalities to be drawn. In fact, contemporary interest in size dependence of production in aquatic organisms derives substantially from the development of a technology that has a unique capacity for sensing and sizing particles over broad areas of distribution in oceanic and freshwater environments. This sampling technology gave rise to a distinctive new step toward understanding the factors controlling aquatic production. Even so, direct spectral observation of the distribution of individual body sizes in large bodies of water poses such methodological difficulties that recourse is still often taken to indirect, less precise estimates derived from average biomass weights and allometric conversion values (Cottingham 1999). Parallel compilations of terrestrial data are even more tentative, perhaps because of the great variation in shape and size of primary producers, and peculiarities in the proliferation and distribution of predators in complex food chains that make the determination of allometric values of population phenomena particularly complex (Vezina 1985; Carbone et al. 1999). At the smaller scales, this is complicated by the fact that the very shape of the terrestrial environment is highly dependent on the vegetation at the base of food chains. We have concluded that it is necessary for us to leave evaluation of the state of development of terrestrial sampling to others with a more special knowledge of the difficulties posed by its various properties.

In aquatic environments, information about the gross characteristics of living organisms and their distribution was initially obtained only from labor-intensive point sampling with nets. Preserved samples were enumerated and sorted—by microscope in the case of the smaller organisms. The relatively long horizontal and vertical scales that needed to be covered meant that much of this sampling tended to smear distributional details in relation to the complex and rapidly changing environmental structures significant to the organization of food chains (Longhurst 1967; Parsons, LeBrasseur, and Fulton 1967; Parsons and LeBrasseur 1970). Thus, over time, the sampling requirements gave rise to increasingly ingenious net-closing devices that allowed more detailed quantitative sampling of particular water masses and layers (Frost and McCrone 1974; Sameoto, Cochrane, and Herman 1977). Finally, however, it has been the particle-counting and -sizing instruments able to process large numbers of successive, small samples that have enabled representative, time-explicit study of variability in both the vertical and horizontal components of the distribution patterns (Parsons, Takahashi, and Hargrave 1984). The necessary technologies have become widely available only during the past quarter century.

This modern technological era in aquatic ecology began with the introduction of the Coulter Counter, a device originally intended for counting blood cells in hospitals, but which in its oceanographic implementation made it possible to enumerate and size a broad class of the smaller particulate constituents contained in successive discrete, virtually point samples of seawater along extensive transects (Sheldon and Parsons 1967). The original equipment distinguished particle sizes ranging from about 3 $\mu$ to 1000 $\mu$ in equivalent spherical diameter. The ambit of instrumental sensitivity has since been expanded into the smaller, bacterial body-size ranges and into larger plankton of the order of 3000 $\mu$, but the range just cited was what was more or less reliable in the early 1970s when it was first applied. It was at that time that R. W. Sheldon took a Coulter Counter to sea for the express purpose of describing the distribution of marine particulate matter over broad areas in both tropical and temperate regions of the Atlantic and Pacific Oceans (Sheldon, Prakash, and Sutcliffe 1972).

It is important to note the automated aspect of this instrumentation. Certainly, a high degree of expert operator intervention was required at the time, and remains necessary today; but it was the automation possible on large oceanographic vessels that permitted so many observations to be made at small spatial and particle-size dimensions over wide ranges of the spatial and temporal scales that characterize marine environments. Automated instrumentation is not unique to particle-size analysis. It is, in fact, one of the key features in the revolutionary development of contemporary oceanography. Instrument development, including automation, has played a major role in describing the size structures of aquatic ecosystems at both small and large spatial scales.

Following the initial samplings of Sheldon and associates (1972, 1973) on the phytoplankton and smaller zooplankton, the intermediate and larger body sizes of the zooplankton were found to be especially difficult to detect and size, partly because of the high packing densities in their swarms, which may occur at variable depths, but often near the seafloor. Sameoto, Cochrane, and Herman (1993) have shown how a combination of technologies, having the principles of Coulter Counters, optical counters, and acoustic reflection techniques as part of their lineage, had to be adapted for surveys of organisms ranging in size from small zooplankton to mesopelagic fishes. With instruments of sufficient sensitivity and sophistication, it has become possible within the last decade to begin the investigation of particle-size aggregations and distributions for most of this range of pelagic organic particles, and to confirm the results using combinations of different sampling techniques (Herman 1992; Sameoto, Jaroszynski, and Fraser 1977; Sameoto, Cochrane, and Herman 1993; Huntley, Zhou, and Nordhausen 1995).

Literature covering the advent of acoustic detection techniques for the larger, fish-size organisms is a speciality in its own right. Acoustic methods for detecting larger sizes of particles was an offshoot of attempts at acoustic submarine detection during wartime (Craig and Forbes 1969), considerably prior to the appearance of the Coulter Counter. In the course of developing this technology, it was found that interpretation of submarine signals was significantly affected by reverberations from massive layers of living organisms present in different water layers (Chapman 1967). There was initially little interest in or capacity to appreciate individual particle size, except as order of magnitude averages affecting the reverberation parameters. However, the high frequency and variability of encounters gradually led to a flourishing of practical methods for better detecting the concentrations and particle sizes of the organisms, some of which appeared to be fish, and eventually to the devising of methods that would enhance the efficiency of sampling methods to check the theoretical calculations. The character and distribution of the masses of the vast pelagic schools of small fish as well as many other classes of organisms that were found in this way had been scarcely known before these developments (Clay and Medwin 1977).

It required considerable ingenuity and many years of effort for acoustic technology to distinguish the sizes and distributions of individual organisms on both a broad and a fine enough scale to be of value to scientific studies of biological energy flow (Foote, Aglen, and Nakken 1986; Dickie and Boudreau 1987). There remain problems in detecting organisms very close to the bottom, and in distinguishing large organisms without air or swim bladders from small organisms that possess these efficient acoustic reflectors. However, a satisfactory methodology has been applied to populations of pelagic fishes occurring in the Great Lakes (Brandt et al. 1991). The technique has empirically revealed patterns in density distribution by body size that enable the checking of observational against theoretical expectations of the nature of density distributions of body sizes of organisms from phytoplankton to fish (Sprules and Goyke 1994). For significant fractions of the pelagic zones of the world oceans and the major freshwater lakes, it has now become possible to enumerate by particle size the biomass densities of the bulk of the major carriers of biological energy.

Two areas of detection of the smallest aquatic organisms remain to be fully developed: for the benthos and for the smallest organisms from bacteria to femtoplankton (Sieburth, Smetacek, and Lenz 1978). The need to measure energy flow in the benthos was recognized following early studies of the dependence of some fisheries on benthic production, and continued through discovery of the high rates of respiration of the benthic layers. Schwinghamer (1981a) was among the first to resort to automated techniques for analyzing benthic samples, particularly the smaller, softbodied organisms. Simple technologies, but often involving massive machinery, had been, and regularly still are used for collecting materials, as well as for subsampling, washing, and filter-sorting the myriad of benthic organisms taken in them. However, the wide range of sizes and shapes encountered in substrates of very different types and the manual labor of handling the massive samples of bottom sediments have continued to be major obstacles to quantification. In such cases, the advent of computer imaging and sizing has offered substantial relief for the enumeration of samples, permitting typing of samples by both size and shape (Ramsay et al. 1997).

Only recently has it seemed possible to automate the sampling for information on the structure of the benthic substrates themselves on a scale that can be related to the size of the habitats of individual small benthic organisms. This has involved the adaptation of ultra-high frequency acoustic techniques in a three-dimensional analysis (Schwinghamer, Guigné, and Siu 1996) that allows detailed examination of both bottom textures and structure on various scales. In combination with data on the contained organisms, these early developments have already demonstrated the effects of changes in substrate on the types of organisms contributing to energy-flow processes in this important sector of the aquatic ecosystem.

Various algorithms have been used to classify the sizes of the smaller planktonic and benthic organisms. At the lower limits, Sieburth, Smetacek, and Lenz (1978) distinguished two groups among the "microplankton" as femtoplankton (0.02–0.2 $\mu$) and picoplankton (0.2–2.0 $\mu$), the latter group evidently including the bacteria. The term "nannoplankton" was reserved for organisms an order of magnitude larger (2.0–20 $\mu$). Sampling methods for these smallest sizes have developed steadily over the past decade, and the results of revisions have had significant effects on overall estimates of biomass and production density (Platt, Lewis, and Geider 1984). A comprehensive appreciation of the technology suitable for measuring them appeared in a volume edited by Platt and Li (1986).

Automation has been slow to develop for constructing biomass spectra for these smallest living particles, which may ultimately need to include the viruses (Fuhrman 1999; Wilhelm and Suttle 1999). In its absence, operational definitions of body size, based on standardized methods of oceanographic sampling and filtering, have proven their usefulness from the point of view of repeatability. Thus, Quiñones (1992) classed as "bacterioplankton" the cells counted by epifluorescence microscope in an aliquot taken from Niskin bottles from particular depths, preserved, stained for fluorescence, and retained on a 0.2 $\mu$m pore-size filter. He applied the terms "microplankton" and "nannoplankton" to the two next-larger body-size groups from the Niskin bottle samples, defined operationally as the living particulate matter that either passed through or was retained by a 35 $\mu$m mesh filter. In such cases, as with bacteria and benthos, both the collection and filtering methods have to be carefully controlled. Therefore, collections are generally still taken in oceanographic sampling bottles of various sizes (generally 5–30 L) at discrete water depths. As with Quiñones' studies, the efficiency of sampling can sometimes be increased by selecting depths according to accompanying vertical profiles of in situ fluorescence. Automation can be employed in certain aspects of enumeration, as with flow cytometry (Li and Dickie 1985).

State-of-the-art biological sampling in aquatic environments, while still undergoing development, is thus now able to sample for both body size and density over a wide body-size range of organisms, from bacteria to large fish. One expects further sampling studies for the smallest sizes to yield new information on the entire spectrum, as methods are further automated or new methods found to enable the sampling of even smaller organisms, such as the viruses, about which so little is yet known. Equally critical, however, is a need for improved estimation procedures for the widespread larger organisms: the birds, seals, and whales (Brodie 1975; Bowen 1997). These groups, in which many segments of society have a strong interest, undoubtedly have strong energetic and predator-prey interactions with the more basal parts of food chains in the sea, but their study requires special methodologies because of their relative numerical rarity, their highly aggregated distributions, and their strong, seasonal, migratory habits. To date, we have little information from them on which to base spectral compilations; so we do not attempt to deal further with them here.

## The Early Evidence

Sheldon's original compilation of the size distributions of plankton in the sea was made on a cruise called Hudson 70, a circumnavigation of the Americas by the research vessel CSS Hudson (Edmonds 1974). On this cruise, he and his associates had ample opportunity to observe the distributions of biomass density (defined here as biomass in $\mu g \times L^{-1}$ projected onto a unit area of the surface of the water body) by body-size classes in a wide variety of oceanographic pelagic situations over significant expanses of the global oceans. Despite the obvious bumps and irregularities manifest in their observations, they were struck by the uniformity of its biomass concentration at the various sites across the range of particle sizes that could be observed directly with the Coulter

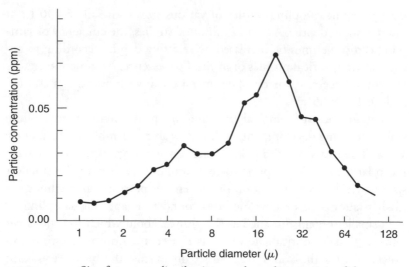

FIGURE 2.1A    Size-frequency distribution; to show the notation of the axes and the least number of data points used to define the form of the distribution. Concentration is by volume. The distributions in figure 2.1B were constructed in this way. (Redrawn from Sheldon et al. 1972.)

Counter. Following the line suggested by his earlier work (Sheldon and Parsons 1967), Sheldon expressed the ordinate of his data plots in logarithmic units of average particle concentration and the abscissa in logarithmic units of body size graduated in octave units of equivalent spherical particle diameter (figure 2.1).

Apparently inspired by Riley's (1963) concept of "potential tuna," Sheldon and associates (1972, 1973) mined the scientific literature to compare their firsthand observations with published accounts. They were able to estimate the relative biomasses of large pelagic organisms, such as tuna and whales, to compare with the biomasses in the phytoplankton and zooplankton body-size ranges. Although unavoidably approximate, the results convincingly showed that the concentration of biological particulate material in the sea is spread with remarkable uniformity across the spectrum of body sizes, from bacteria to whales.

Initial response to this generalization ranged from ennui ("Oh yes, merely the Eltonian pyramid, expressed in slightly different terms") to astonishment ("How can we account for this remarkable uniformity!"). The theoreticians were not long in responding to the implicit challenge.

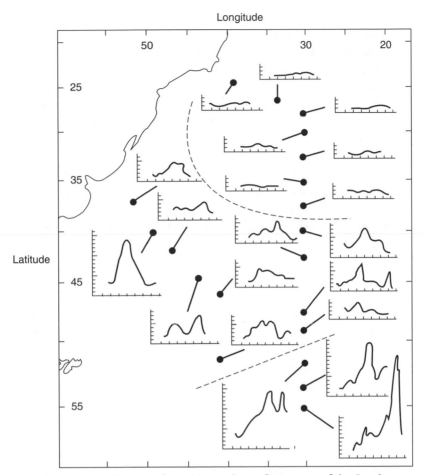

FIGURE 2.1B    Distribution of particles in the surface waters of the South Atlantic. The broken line in the north represents the limit of the subtropical water as indicated by the particle spectra. The broken line in the south represents the approximate position of the Antarctic Convergence. (Redrawn from Sheldon et al. 1972.)

Because we deal with theoretical aspects of the spectrum at greater length in chapters 4 and 5, for the moment we note only that the basic analytical considerations first explicitly identified by Kerr (1974) figure in all subsequent theoretical analyses of size distributions in aquatic ecosystems. Without exception, these are built on the perception that the organization of aquatic ecosystems derives importantly from energy transmission through predator-prey interactions in which large organisms, on average, consume smaller organisms as prey, and that all organisms are subject to the familiar allometric decay of weight-specific metabolic rate as a function of increasing body size.

Platt and Denman (1977, 1978), while recognizing predation as a major means of transmission of energy from small to larger body sizes, were more concerned with data on the major groupings of small phytoplanktonic and bacteria-size organisms at the base of the food chains among which particulate feeding may be less common. Their mathematical analyses envisioned the spectrum as a continuum and so directed attention toward calculating the efficiencies of transfer of energy between the adjacent size classes that were suggested by their differential equations. Despite this basic difference in underlying model from that of Sheldon and associates, the important result of the initial examinations was common perception that, given even simple assumptions respecting the rates of energy uptake and flow through aquatic food chains, expected trends in the concentration of living materials over the broad spectrum of organism sizes in the oceans were in logical accord with the generalization of observations originally made by Sheldon and his colleagues.

This initial evidence from offshore marine ecosystems, together with its plausible if rudimentary theoretical underpinnings, moved others to explore the phenomena in a variety of aquatic environments. Among the earliest was Schwinghamer (1981b) who made observations on the benthos, including bacteria with average equivalent spherical diameter of less than 1.0 $\mu$m, through the larger, microscopic infauna to the macroscopic epifaunal invertebrates. It was he who first drew attention to the fact that over the broad range of sizes in his samples there were distinct dome-shaped subgroupings of the total biomass into body-size categories. Three subgroups were recognizable, with near-zero biomass density gaps at body sizes between them.

On the basis of this perceived internal structure, Schwinghamer (1981b) suggested that, in addition to the primary organizing effects of predator-prey food-web interactions in the ecosystem (Smith 1976), the basic size groupings might reflect, at the microhabitat level, the secondary ecological organizing effects of the particle sizes of the substrates that compose the environment in and on which benthic organisms are found. Thus, he divided the whole biomass sampled by the grabs into three size-graded subgroups: the smallest living organisms, primarily the bacteria, that live on the surfaces of the finest particles of their substrate; a larger group of microalgae and meiofauna that live in the interstices among the particles; and a largest, epifaunal group, which lives on the surface of the substrate. Schwinghamer compiled his results in much the same way as Sheldon, Prakash, and Sutcliffe (1972), except that, since he was not using a numerical particle-size counter, he represented the ordinate of his plots in logarithmic intervals of biomass concentration.

Schwinghamer's first observations were confined to sampling of coastal habitats. Subsequent sampling broadened the types of environment examined. Schwinghamer (1986) showed that the same basic size categories that appeared in the samples taken from inshore and intertidal benthic communities were also characteristic of the benthic biomass distributions sampled from the continental shelves and from the abyssal plain. We describe his results in more detail in the next chapter in association with other "analytical" approaches.

At about the same time, Sprules, Casselman, and Shuter (1983) undertook analysis of pelagic particles in 37 Ontario Lakes, using the methods that had been suggested by Sheldon and Schwinghamer. Their results showed that over a body-size range of nearly four orders of magnitude from about $2\ \mu$ to $3000\ \mu$, the biomass spectrum of phytoplankton and zooplankton organisms was characterized by definite domes in the size distribution. Two distinct domes of biomass density occurred in the range of plankton body sizes in all of the lakes. These domes were clearly defined by three size ranges in which there was zero or near-zero biomass. The gaps in the distribution were judged not to be artifactual, hence to represent apparent limitations to the physical size sustainable by planktonic organisms in freshwater. Sprules, Casselman, and Shuter (1983) compared the ratio of biomass densities of phytoplankton and zooplankton in the lakes and the sea and found them to be similar. They also deduced from more general information on fish biomass density that the overall spectral slope they observed in the plankton applied over the full range of pelagic particle sizes present in the lakes. That is, marine and freshwater spectra seemed to show common characteristic shapes and slopes, possibly indicative of dynamic controls common to these two types of living aquatic energy-flow

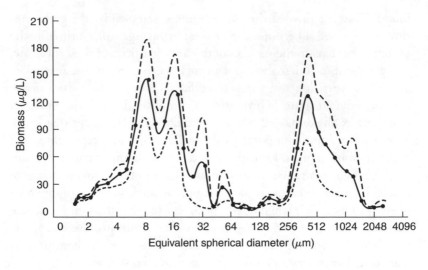

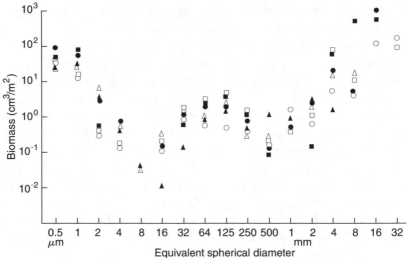

FIGURE 2.2  Comparison of particle-size distributions in freshwater lakes and the marine benthos, indicating the complementarity between peaks and troughs. Upper panel shows average particle-size distribution for 26 lakes together with 95% confidence interval. Lower panel shows biomass distributions at six sampling stations in Atlantic Canada. Note that the concentration indices (ordinate) differ, but the particle sizes are plotted at approximately the same scale (abscissa). (Upper panel redrawn from Sprules et al. 1983; lower panel redrawn from Schwinghamer 1981b.)

systems. Strayer (1986), on the basis of limited sampling of the benthos of a small lake, pointed to some situations that are exceptions to this generality, especially among the smallest zooplankton.

One further significant step in appreciating the possible generality of an integral spectrum, having at the same time significant internal structure, came from Sprules and associates comparing their sampling results with those of Schwinghamer. There were distinct differences between the average body sizes of the major trophic groups in Sprules's pelagic sampling and the average body sizes that had appeared in Schwinghamer's benthic sampling. In fact, the domelike structuring in the pelagic and benthic spectral size distributions seemed to form complementary series, in which the peaks in the pelagic distribution fitted the gaps in the benthic (figure 2.2). Sprules and associates noted that the patterning of this interdigitation offered support for Schwinghamer's hypothesis (1981b, 1986) that in these relatively shallow-water environments the size compositions indicated a basic coupling of the pelagic and benthic subsystems into a unified ecosystem, primarily through a linkage of their trophic energy flows.

## Initial Approaches to the Dynamics of the Spectrum

As a result of these observations, comparisons, and interpretations, the stage was set for more decided attempts to understand the dynamics underlying what was emerging as a remarkably stable and ubiquitous set of characteristics. The overall or integral spectrum of biomass density distributions seemed to have a common pattern in all aquatic environments. Information on seasonal variability was lacking, but this might be expected to appear with time. What was now required, as a basis for scientific testing of the appropriate degree of aggregation of observations against ideas about mechanism, was a stronger postulation of formal relationships among the main structural features and their possible dynamics.

### Observations on Partial Spectra

Platt and Denman (1977, 1978) had appreciated that the overall slope and position of the integral biomass spectrum could offer specific and practically useful measures of total energy flow in ecosystems. Sheldon and associates (1972, 1977) had attempted to demonstrate this aspect of the utility of their ideas by invoking the characteristic turnover times of different body sizes in prey-predator interactions as a means of interpreting the observed transmission of biomass between plankton and fish in relation to potential fish yield. However, the very general form of their argument offered little in the way of new specific insights into the mechanisms involved in any known situation. On the other hand, interest aroused by Platt and Denman's (1977, 1978) explicit mathematical formulations called for testing of hypotheses about spectral dynamics by fitting data to theory with much greater precision. For these purposes the original description of the spectral slope as "flat" by Sheldon and his associates was not considered helpful.

Kerr (1974) attempted greater precision, and his results implied that better measurements of the internal production processes would help establish the relative roles of physiological constraints and ecological considerations related to predator-prey body-size ratios in the food webs. As he pointed out, the predator-prey body-size ratio must dictate at least a horizontal or abscissal spacing of the biomass elements that determine the overall biomass density spectral slope.

Platt and Denman's extension of this logical enquiry was based on a more sophisticated mathematical analysis designed to explain estimates of the general level of biomass and slope of fittings over the phytoplankton and zooplankton body sizes. Their theoretical analyses were based on a perception of total energy flow as a continuous process through contiguous size groups with an intertwining of energy flows both up and down the body-size spectrum. These might take place through possibly complex feeding relationships that were juxtaposed to the regeneration processes in which the bacteria play such a prominent role. They also drew attention to the flows of generative materials from large to small body sizes during reproduction, especially among the larger aquatic organisms. These considerations suggested to them, and later to Silvert and Platt (1980), the need for a generalized characterization of material and energy movements from compartment to compartment of the spectrum. The implication of this approach was that a particular average transfer value, embodied in the slope of the spectrum, would offer an opportunity to use spectral parameters for improved predictions of overall ecosystem production: fish production might well be predictable from measurement of phytoplankton production.

A logical extension of the purely statistical aspects of fitting a continuous curve to generalized data over the entire observed body-size range has been given more recently by Vidondo and associates (1997), where, however, the connection of the description with its possible generating mechanisms has been virtually lost. A somewhat similar theoretical analysis of the consequences of different types of statistical fittings has been undertaken by Han and Straškraba (personal communication, June 1998), who point out how the selection of different degrees of aggregation of elements of observation may affect the main estimates. A more promising practical development of the continuous spectrum is that described by Zhou and Huntley (1997), who take explicit account of the internal processes of growth and reproduction in relation to metabolic and predation dissipation to understand the production relationships among discrete small water masses. The dominant concept of all of these developments is that of an underlying continuous net flow of energy and particles up the size spectrum. This should result in uniform straight lines of biomass with body size, having possibly varying slopes, or exhibiting simple polymodal curves. As discussed in succeeding chapters, it produces an image of the spectrum that is in sharp contrast to the predator-prey models of Sheldon and Kerr. The significance of fittings of the two different models to the testing of theory by empirical data was not initially appreciated.

Platt and Denman's approach may be said to have given expression to the particular interest in establishing an index of general rate of energy flow for the overall system, rather than attempting any more detailed explication of the organization of elements within it. As a result, they directed attention to the meaning of the various constants whose measurement was called for by the solutions to their allometric equations for the integral spectrum. The danger of this search for generality is the temptation to find it at the expense of perceiving the significance of the more detailed structure. We return to this important question at later stages in our analyses, in relation to phenomena that arise at different scales of observation. The differences assume practical importance in applying spectral interpretations to management of natural resource production.

For its mathematical convenience, Platt and Denman also introduced the "normalized" spectrum, in which biomass density within each size group is expressed as a function of the width of the size group (see, for example, figure 2.3). This calculation makes the slope and intercept of the normalized spectral representation independent of the sometimes almost unavoidable differences in the dimensions of the body-size classifications in a given spectrum. In its formal aspects, this methodology merely converts a horizontal spectrum of the unnormalized Sheldon spectrum, with a near-zero slope, into a normalized spectrum with a slope near $-1.0$. But the scale changes accompanying it offer the distinct advantage of a generalized representation of the spectrum for comparative purposes. Platt and Denman's analysis thus directed attention particularly to how properties of the integral spectrum might be used to characterize and compare different aquatic biological community dynamics, a need for which they had convincingly argued in their perceptive earlier work on the spectra of variability in aquatic ecosystems (Platt and Denman 1975).

A major contribution of the 1977–1978 analyses of Platt and Denman was their recognition of the central importance of the allometry of metabolism of the body size of individual organisms in determining the overall slope of the biomass density spectrum (see also Platt 1985). This was formulated precisely in thermodynamic terms in their 1978 account, where they sought to interpret the constants that arose from the mathematical integration of the equilibrium spectral equations in terms related to the rate of turnover in the biological community. The general form of the solution they adopted was, in fact, strongly parallel with the much less precise interpretations that had been undertaken by Sheldon, Prakash, and Sutcliffe (1972). Both recognized the central importance of the fact that the sustained and stable biomass densities observed must result from a state of balance between the growth and dissipation processes of the individuals composing the biomass.

Platt and Denman's more precise fitting of their spectral equations to data relied on prior estimates of the parameters describing the weight dependence of metabolism and turnover time given in the well-known general paper of Fenchel (1974). In this connection it is worth pointing out that Fenchel gave estimates for two vital dynamic parameters: metabolic rate and intrinsic rate of increase, for organisms from bacteria to large mammals. Platt and Denman (1978) adopted both of them for their equations. First of all, they accepted the metabolic rate parameter, which had been widely examined and agreed on. Secondly, they equated Fenchel's estimate of intrinsic rate of increase with their inverse of the turnover time. As pointed out by Boudreau and Dickie (1992), the validity of their estimate of a specific spectral slope for the biomass spectrum thus depends more heavily on judgments about the appropriateness of equating Fenchel's estimate of the intrinsic rate of increase with the inverse of the observed population turnover time than it does on any reservations about the statistical fittings of the empirical data. Intrinsic rate of increase is a theoretical population value that seems unlikely ever to be realized in an actual ecosystem. By contrast, turnover time is an empirical value derived from estimates of average population growth rates under normal energy-limited, ecological conditions. We prefer to reserve judgment on the most appropriate benchmark for spectral interpretation pending discussions of the significance of various allometric fittings of production data to different scales of aggregation of organisms, as in chapter 3.

Platt and Denman's calculations involving these and several other less important parameters resulted in an estimate of the normalized spectral slope of $-1.22$, a value that they recognized as needing refinement, but that, in subsequent papers, they and their associates tended to adopt as a kind of benchmark against which to judge the validity of both sampling data and their dynamic interpretations. They recognized, however, that transfer conditions throughout the spectrum were unlikely to be completely constant and that more needed to be learned through study of the properties of spectra over a wider band of frequencies and body sizes in a greater variety of environments. This inquiry was briefly taken up by Platt, Lewis, and Geider (1984), where the new data from the North Pacific Central Gyre published by Beers, Reid, and Stewart (1982) could be included. These data, based on microscopic observations of a fuller range of plankton sizes, were thus also free of criticisms that had been raised about the limitations of Coulter Counters for detecting the full size-range of particles, and of concerns about the instrument's inability to distinguish between dead and living particles.

Their results suggested that there was high seasonal stability of the particle-size distribution in this large, strongly oligotrophic area of the Pacific Ocean. Their statistical fittings of the improved data on the concentrations of particle sizes between $2\ \mu$ and $232\ \mu$ were taken to support Platt and Denman's earlier (1978) prediction of a slope of the normalized spectrum of the order of $-1.22$. In their judgment, slopes of this order were not adequately comprehended by the earlier description of the spectrum as "flat" by Sheldon and associates, with its implied normalized slope of only $-1.0$.

However, Platt, Lewis, and Geider (1984) also appreciated that because of the relatively small size-range of observed particles, the data were liable to a certain subjectivity of interpretation. In particular, they noted that extrapolation of the Pacific Gyre microplankton spectrum to the picoplankton sizes that had been missed earlier yielded a lower overall slope of the spectrum. The new estimate was between $-1.11$ and $-1.14$. They utilized a combination of this and their earlier distributions as a basis for calculating total plankton community respiration. That is, they proposed the spectrum as an indirect but more reliable method for checking conventional carbon assimilation estimates of gross primary production in the sea by use of the spectrum. It was clear, however, that still more information was needed over a wider range of sizes. Even the small additional data on the previously undetected picoplankton body-size range changed the estimates of overall planktonic community respiration by a factor of three.

The effects of utilizing a particle-size spectrum of the plankton on understanding differences in basic production were further examined by Moloney and Field (1985) and Moloney, Field, and Lucas (1991). On the basis of their observations in three different types of oceanographic production areas, represented in the upwelling and more stable coastal shelf areas off South Africa, they developed size-based biological models and used the parameters to calculate differences in the productivity of pelagic fish food chains in relation to the observed differences in nutrient concentrations. Their models demonstrated that the effects of input nutrient concentration are mediated through concomitant effects in the whole of the particle size distribution in ecosystems. They deduced that such particle size-distribution effects seemed to be initiated in the phytoplankton, but their models further showed that consequent differences in productivity were inadequately explained unless the explanation included a recognition of the promulgation of these size effects throughout the communities of pelagic organisms.

It was Sprules and Munawar (1986) who undertook the first extensive analysis of empirical data on the planktonic phases of the particle-size spectrum of a variety of environments. By that time, the data from Beers and associates for the oligotrophic North Pacific Central Gyre had been extended by Rodríguez and Mullin (1986) to include new observations on the macroplankton; and the seasonally extensive data of Sprules, Casselman, and Shuter (1983) on small eutrophic inland lakes of Ontario had been supplemented by data from several of the Laurentian Great Lakes of North America. Sprules and Munawar (1986) thus made available data on micro- and macroplankton over a minimum of eight octaves of body size in their $\log_2$ body-size classifications of biomass density in seven trophically different environments.

Their statistics of the normalized distributions revealed a gradation in slope of the spectrum from $-0.98$ to $-1.16$, which they tentatively concluded may possibly be related to differences in the nutrient supplies to the ecosystems studied (figure 2.3). The steepest slope $(-1.16)$ appeared in the newly amplified data for the highly oligotrophic North Pacific Central Gyre for which Platt and Denman had made calculations. The slope for this oligotrophic area was in greatest contrast with the relatively "flat" spectra for the four most productive of the eutrophic Great Lakes series. These included the small but highly dynamic and productive Lake St. Clair in the chain of the Great Lakes between Lake Huron and Lake Erie, together with the three smallest and most productive of the Great Lakes themselves, where slope values were close to $-1.0$.

Utilizing models of Walters, Park, and Koonce (1980), Sprules and Munawar pointed out that zooplankton abundance would be expected to vary two to three times more than nannoplankton abundance in the face of changes in nutrient supply of the order indicated by their data. This possibility was later underlined more emphatically by Ahrens and Peters (1991a), who found, in a series of small lakes, that the zooplankton biomass distributions often had dome peaks that were higher than those of the phytoplankton (i.e., the normalized slope was significantly less than $-1.0$). We discuss their observations in more detail in chapter 3. However, as Sprules and Munawar (1986) pointed out, the relatively high variation in zooplankton biomass densities in relation to phytoplankton densities made it somewhat premature for simple extrapolation of plankton production data to be considered useful for predicting fish yields. A similar conclusion may be inferred from the extensive compendium of International Biological Program (IBP) data by

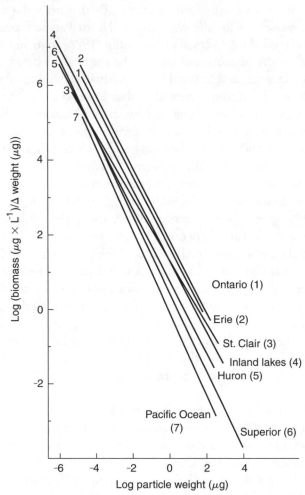

FIGURE 2.3 Normalized biomass spectra for six freshwater study sites and the North Pacific Central Gyre. Numerals are to aid in identifying both extremes of the lines. Details of the fittings are provided in the original reference. (Redrawn from Sprules and Munawar 1986.)

Cyr and Peters (1996). They found no significant slopes of the biomass spectra, but concluded that in small to intermediate-size lakes, phytoplankton were better predictors of fish biomass than were zooplankton. Unfortunately, with the wide scatter of points, it is uncertain how to differentiate methodological from main effects in their conclusions.

A further set of pelagic plankton observations from various areas of the eastern and western Atlantic areas has been made by Quiñones and associates (1992, 1994). In his observation series he made considerable efforts to include data on the bacterial sizes, along with estimates for the more familiar ranges of phytoplankton and zooplankton. The results, when plotted in terms of biomass density of live weights of individual organisms, showed slopes that vary but little from $-1.0$, a value close to that implied by the original summation of Sheldon, Prakash, and Sutcliffe (1972). In an apparent effort to examine the possible correspondence between his representation and the 1978 theory of Platt and Denman, Quiñones recalculated his biomass spectra in terms of equivalent volumes of carbon, a move justified by a perception that metabolic rate, like primary production measures, may be more clearly related to carbon content than to the mass of living tissue. The results yield generally slightly steeper slopes, closer to those of Platt and Denman. However, the difference, while of uncertain statistical significance, also depends on the questionable accuracy of the carbon equivalents adopted. We consider other, more important aspects of his analysis in chapter 3.

At about the same time as these attempts to relate the empirical data to the theoretical underpinnings of the continuous flow models, there were a number of attempts to use the growing information to develop theory that would support the extrapolation of the trends to the full biomass spectrum. It was recognized that the biomass spectrum, while apparently capturing the main characteristics of the biomass distribution in observed systems, was essentially a stable model. That is, it was not a time-dependent model, and thus had limitations as a device for deducing the dynamics of ecosystems or for extrapolating to the full system. Silvert and Platt (1978, 1980) proposed a time-dependent modification of the Platt and Denman continuous model by taking explicit account of certain additional general aspects of both predation and growth transfers that had not been made clear in the original accounts. Their new approach had the important result of pointing the way to possible empirical tests of alternative theories of energy transmission by observations in time. This possibility was first appreciated by Vezina (1986), who designed and carried out a series of test experiments. We consider Silvert and Platt's model and compare its implications with the results of Vezina's tests in relation to the internal structure of the spectrum in chapter 3, because its significance is strongly connected with assessing the relative suitabilities of the continuous model and the predator-prey discontinuous model as alternative means of estimating and interpreting spectral parameters.

Platt and Silvert (1981) also considered the theoretical implications of dimensional analysis for interpreting the slope of the biomass spectrum. Their conclusion that one might expect a small difference in slope, although one perhaps not empirically perceptible, between the spectra for terrestrial and aquatic systems seems to us to be based on a confusion of the effects of individual and population properties on allometric coefficients and the manner in which they would be manifested in aggregations of organisms.

Borgmann (1983, 1987) also considered the problem of modeling the integral biomass spectrum in relation to its possible generating mechanisms. He deduced that the straight lines fitted to the continuous models of Platt and Denman implied an increase in transfer efficiency through the food chain, and through this effect imposed a steeper slope on the spectrum than was justified by empirical data. He proposed an alternative theoretical development that assumed a constant transfer efficiency between trophic steps. His results suggested that the slope of the normalized biomass density spectrum should be $-1.02$ to $-1.04$, corresponding to values established empirically by both his own plankton data from Lake Ontario and the original Sheldon plots. His theory assumed, however, that all prey production was consumed by predators, an approximation that was unacceptable to some workers, despite the growing evidence for generally lower slopes in the more productive systems.

## Extensions of the Spectra to Fish-Size Particles

Fisheries data are traditionally compiled in terms of total population abundances, related to economic yields. This makes it difficult to compare them directly with plankton data, which are customarily compiled in terms of population densities. In the late 1980s, however, two studies of the biomass spectrum found that allowances could usefully be made for the missing details on sampling area or volume in the fisheries statistics. This made available additional important empirical data related to integral spectra. Overall slopes of the order of $-1.04$ were found by Minns and associates (1987) for the highly productive Bay of Quinte on the northeast side of Lake Ontario. Similar slopes were also found by Boudreau and Dickie (1989) in their analyses of data for a number of freshwater and marine areas. The uniformity of biomass density estimates between integral spectra for various systems, and the smooth overall slopes for the resulting more complete spectral representations, provided some of the assurance of generality that had been called for both by Sprules and by Platt and Denman.

In addition, Boudreau and Dickie (1989) showed that changes in biomass and production densities within individual cohorts of the multiaged populations of commercially exploited fish species exhibited relations to body size that were consistent with interpretations of the spectrum that had been developed from data on the smaller, shorter-lived planktonic and benthic species. The evidence pointed to the presence of a common, pervasive set of factors controlling metabolic rate and trophic energy uptake as the basis for the regularity of both the spectral parameters and the structures they described. Analyses and comparisons of the empirical data from the variety of sources left little doubt that the underlying dynamics of the spectrum including the fishes were indeed reflecting the controlling influence of the conditions of energy supply, dissipation, and transfer within whole ecosystems.

Given the empirical evidence that measurements from a partial spectrum of an ecosystem may fail to reflect properties of the whole spectrum, we end this chapter by looking at a review of several integral spectra undertaken by Boudreau and Dickie (1992). They compiled suites of biological data on plankton, benthos, and fish from ecosystems in various parts of the world. Some of the data sets represent the biomass densities from major fisheries-related ecosystems. As might be expected, some of the data lacked the kind of detail available from Coulter Counter–type sampling. However, the authors concluded that a compilation of body-size data in $\log_{10}$ intervals would permit sufficiently detailed construction of "complete" spectra. They were then able to compare the biological data with characteristics of the physical environments in which the data were developed in more detail than had been possible for Sprules and Munawar (1986). The results allow us to venture further toward a comparative ecology based on properties of the integral spectrum.

Details of the sampling information and conversion factors used by Boudreau and Dickie are available in the original paper. Data were assembled on phytoplankton, zooplankton, and fish populations for four marine and freshwater pelagic ecosystems. Benthic data were also available for three additional marine systems. Domes of size distribution of biomass could be plotted for several of these sampled areas, and the detailed data for some of them are discussed in chapter 7. For consideration of integral spectra, however, the importance of these details is that they allow a comparison of the height and position of domes in the biomass distributions of fish species with data series where average biomass and body size was known but the detail was either not preserved or not determined in the original observations. The empirical relationship between the average biomass density of the average body size and corresponding peaks of spectral domes was used to complete the data series for two of the fisheries systems, allowing their inclusion in the comparisons of integral spectra.

The resulting calculations of annual biomass density for phytoplankton, zooplankton, benthos, and fish are displayed in figure 2.4. There are a minimum of three points per ecosystem for six systems. A seventh system, the Pacific Gyre, is represented by only the two points for phytoplankton and zooplankton since few fish were taken by the sampling gear. But in three of the six "completely" sampled ecosystems the benthos data are given. Comparative interpretation of the integral spectra of these systems in relation to knowledge of the physical oceanographic properties associated with energy supplies to their environments supports our growing expectations of the ecological value of the integral spectral methodology.

The most striking result of the compilation of average values in figure 2.4 is the apparent parallelism of the slopes for seven major ecosystems in quite different parts of the world. This parallelism depends heavily on the relation between the biomass densities of phytoplankton and fish. The intermediate points for zooplankton or for zooplankton and benthos together are more variable. The findings thus support the conclusions of Sprules and Munawar (1986) about the relative variability of zooplankton density, at the same time suggesting that it is important to be able to extend the range of the empirically determined biomass density patterns into the body sizes of commercially exploited fishes.

The average biomass concentration per system (intercepts) varied twice as much among these seven systems as in the suite of systems examined by Sprules and Munawar (1986) where the range of the maximum to minimum biomass density between systems varied less than

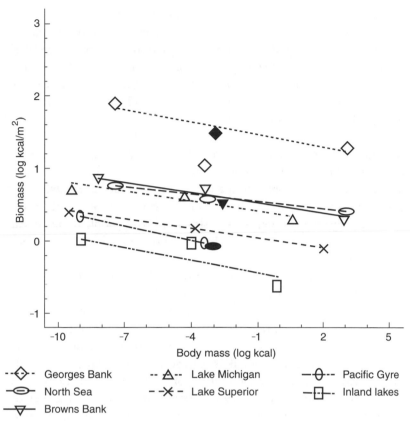

FIGURE 2.4    Normalized biomass spectra for pelagic ecosystems using summary mean points for phytoplankton, zooplankton, benthos, and fish groups. Benthos points are represented by solid symbols. The linear regression for each environment is calculated using the sum of zooplankton and benthos as the intermediate point. (Redrawn from Boudreau and Dickie 1992.)

one order of magnitude. In the Boudreau-Dickie study, the range of biomass densities between the richest and most productive and the poorest system was two orders of magnitude. The high is represented by Georges Bank, a major fishing bank off the east coast of North America. The low is represented by the aggregated data for the group of 37 Laurentian inland lakes, compiled by Sprules and Munawar.

The three ecosystems for which benthic data were available are Georges Bank, the North Sea, and the Browns Bank area of the Scotian Shelf, all historically important fishing areas. These three systems also displayed major differences in the fittings of the biomass densities for the intermediate spectral points. These differences turn out to be particularly instructive in the search for explanation of the overall regularities that emerge from the comparisons of integral spectra. We therefore examine the sampling results summarized in figure 2.4 in more detail.

In the case of the highly productive Georges Bank, the point representing the average benthic density fitted well on a spectral line drawn through the three points: phytoplankton, benthos, fish. However, the zooplankton concentration was only 40% of that of the benthos. In the case of the North Sea the opposite situation appeared. Here the spectral line fitted to the phytoplankton and fish also included the zooplankton concentration, while the benthic density was only 20% of it. The Browns Bank data showed an intermediate situation, with zooplankton appearing a little above a general spectral line and the benthic concentration a little below it.

These differences in the fittings of the major biological components of these systems are of particular interest when it is realized that they correspond with what might be expected from the differences in the physical oceanographic regimes that control the conditions of biological energy supply and circulation in the ecosystems. As Boudreau and Dickie point out, Georges Bank is at the center of a strong gyre circulation caused by the very high amplitude of the Bay of Fundy tides on the region (Loder and Greenberg 1986). As a result of this tidally generated circulation, the waters over the bank are vertically thoroughly mixed and fertile at all seasons of the year, a fact that may account for its especially high biological productivity. It is in keeping with this continual mixing from top to bottom over the whole of the annual seasonal climatic oscillations that the middle of the spectral line appears to be fitted by the average point for the benthic biomass concentration alone.

By contrast, the North Sea is a water body that is vertically mixed in the winter season, but is typically stratified during the late spring, summer, and early autumn. In such systems a major part of the annual production is dependent on a recycling of nutrients within the upper water layers, from which there is a variable rain of biological materials to the benthos from late spring to early autumn. This annual stratification and isolation of the surface waters, with a consequent seasonal nutrient depletion, is followed by an influx of nutrients from the bottom and a resurgence of production, beginning at the time of the fall mixing. The high production continues throughout the winter and again in early spring, as long as vertical mixing is maintained under conditions of sufficient sunlight. This seasonal stratification seems consistent with both the generally lower average biomass density of the North Sea, and with the fact that in the spectral plot the zooplankton, rather than the benthos, showed the most important relation to the trend of general biomass concentration. In contrast with the Georges Bank situation, in the North Sea the zooplankton rather than benthos are in the only position to act as a channel through which trophic energy can flow from primary producers to fish during the longest productive period of the year.

Browns Bank is intermediate in oceanographic type between the other two. It is a relatively smaller bank on the Scotian Shelf to the east of Georges Bank, separated from it by a deep, cross-shelf channel. However, while it is oceanographically separate from Georges Bank, it is still under the influence of the Bay of Fundy tides. Tidal effects on its vertical circulation are evident in samples of the water column taken during periods when there is relatively little disturbance by winds. However, the region of this smaller bank is also completely open to the full force of easterly gales. Particularly in fall and winter, the oceanic winds sweep away the circulation patterns established under tidal influence, replacing them with stratified water from the more easterly parts of the shelf (Smith 1989). In keeping with this situation, the annual average biomass density of this western shelf area is comparable to that of the North Sea. But in this case the benthic and zooplankton biomass densities appeared to be equally important as channels for the trophic energy flow to higher levels of the food chains. The spectral slope on Browns Bank is established by the average biomass densities for all four of the component biomass domes.

In the light of what is known about these oceanographic systems, the differences between the levels of the spectral lines appear to confirm the sensitivity of their intercepts to differences in conditions determining the rate of supply and circulation of nutrients. Average standing stocks, biomass densities, or nutrient concentrations, by themselves, cannot provide accurate measures of productivity differences, which are related to the turnover rates of the populations rather than their biomasses. However, turnover rate is itself a function of metabolic rate, which is closely indexed by body size. The horizontal positions of points representing the trophic groups in the different spectral lines are at nearly the

same average body sizes, which implies that the vertical distances between the spectral lines may be a better reflection of the relative biological productivities of the regions than, for example, average nutrient
concentrations. The results support the conclusion that differences in
realized biological productivity of each of these marine areas are the result of a measurable interaction of the potential nutrient energy supply,
its circulation in the system, and the expression of this availability to the
body-size distributions of average organism densities integrated over the
production seasons.

The apparent parallelism of the low spectral slopes further suggests
that given an energy supply, its resulting distribution throughout living
communities is a common function of the efficiency of its uptake and
the organization of its transmission by aquatic organisms throughout
the food chains. That is, the fact that average body sizes of the corresponding major functional groups are much the same in all the ecosystems indicates a strong parallelism in the organization of their food
chains. The extent to which this functional parallelism could be utilized
in a more detailed comparative ecology depends on a closer study of the
internal dynamics of the system, and further development of the theory
on which more detailed models could be built.

The stability of the integral pattern in this Boudreau and Dickie
(1992) study, in association with spectra from the earlier studies, justi-
fies exploratory generalizations of the parameters describing spectral positions and slopes. Such generalizations were invoked by Platt, Lewis,
and Geider (1984) as potentially useful diagnostic tools for comparing
various systems. We will return to a reconsideration of these possibilities
in chapter 9. Meanwhile, the preliminary evidence that an integral spectrum may not fully reflect the internal dynamics of ecosystems requires
that these generalizations be juxtaposed with the apparently strong patterns of stability of the internal structure. General system theory would
lead us to expect in principle that large-scale effects, such as the environmental effects deduced from comparative study of the integral parameters, could not be overridden by small-scale factors acting within
the structured ecosystem. However, the validity of applying the theory
on which this principle is based needs to be verified by information on
the scales of the mechanisms that actually underlie the structure. In any
case, knowing the extent to which the integral spectrum depends on the
internal structures revealed by an analytic spectrum will strengthen our

ability to extract practically useful production information from spectral data. We begin a review of the required, more detailed studies in the next chapter.

## Summary

We have concluded that, with few exceptions, a spectral display of biomass densities over the whole range of body sizes of the organisms that can be sampled in natural aquatic communities reveals a remarkable stability of pattern. The parameters of intercept and slope of this integral spectrum effectively describe the simple straight lines that have been fitted to the data. Furthermore, comparisons among spectra for a number of different aquatic environments suggest that the major difference between areas is in the level of the intercept of the spectral lines, and may reflect differences in basic nutrient supply and availability. The slopes for integral spectra among different areas seem to change little. That is, a simple comparative ecology suggests that the integral biomass spectrum might be used to improve our predictions of the influence of general environmental factors on the productivity of aquatic ecosystems.

However, not all observed ecosystems have been found to perfectly fit this generalization. Furthermore, there may be significant mechanisms of energy uptake underlying this persistent internal patterning of the distribution of biomass density with body size that are not effectively apprehended in the simple, integral fittings. Thus the clear advantages of being able to characterize whole ecosystems by integral methods may not yet be realizable. The predictive power of the construction of an integral spectrum depends on the strength of links between the environmental factors and the mechanisms through which the effects are mediated. We begin a study of the more detailed structure of the spectrum and of the mechanisms that underlie it in the next chapter.

∾

# CHAPTER 3

# The Detailed Structure of the Spectrum

The possibility that biomass spectra in quite different parts of the world have more or less parallel slopes implies that the spectrum within any one ecosystem must show strong stability. The existence of a general, stable spectrum would, in its turn, offer the possibility that the whole spectrum for a given region or system might be reconstructed from intensive field sampling of a relatively small part of it. This would permit extrapolation, for example, from plankton to the production of larger organisms, such as fish, which has been one of the chief objectives of oceanographic study. If this were appropriate, the biomass density spectrum would offer stronger possibilities of comparison, generalization, and prediction than have heretofore been available through simple correlations between ecosystem components, such as the observed correlations between certain measures of primary production and fish production (Nixon et al. 1986). However, as pointed out by Sprules and Munawar (1986), and found by Boudreau and Dickie (1992), variability in zooplankton or between zooplankton and benthos may be of a different order than that within the pelagic phytoplankton. This indicates that a better understanding of factors that control the internal structure of the spectrum in relation to its integral representation is needed before the possibilities of such application can be assessed.[1]

In any case, knowledge of internal structure would open the prospect that persistent details at this level might be invoked to help broaden an understanding of ways in which the organization of ecosystems is established and maintained, something that cannot be expected to emerge from the dynamics of a purely integral description. An understanding of the kinds of factors that could control a stable ecosystem organization has been missing from theories of whole ecosystem dynamics, although such understanding is essential to improved management application of the knowledge of production.

In this chapter we review development of the line of evidence that indicates that the stable internal structure that appears to be characteristic of the biomass density spectrum depends on mechanisms that control both the organization of the system and the rate of the energy flow that is measured as production in it. As we shall see, the new knowledge of the internal structural features interacts with our increasing understanding of general biological theory to become the basis for a new theoretical model of the biomass spectrum. We begin to elaborate this theory in chapter 4.

Before considering the details, however, it is worth noting that failures in the management of ecosystems, particularly in relation to fisheries, have been reflected at least as much in unanticipated changes of balance and interaction among sizes and species of organisms within the system as in the changes in the overall abundance of target species that have been the principal focus of management research (Nikolsky 1963). Empirical verification of spectral theory models that are both more comprehensive and more detailed than existing fisheries models shows the new potential for management to take into account these previously inexplicable imbalances. More precise description, then, helps to establish the extent to which given changes depend on internal dynamics, external environmental perturbation, or intensive exploitation. The relation of various internal and external influences to spectral phenomena will be discussed, beginning in chapter 7 where we can take advantage of the extensive review of both empirically detected properties and underlying ecosystem theory.

## The Organization of Trophic Dynamics Among Populations of Organisms

The observations of both Schwinghamer (1981b) and Sprules, Casselman, and Shuter (1983) confirmed the original observations of Sheldon, Prakash, and Sutcliffe (1972) that an overall or integral spectrum with characterizable properties related to system productivity could be discerned in the otherwise highly variable data. However, they were also the first to point to distinctive internal patterns denoting an internal organization with significant stability that could be seen within this over-all framework. This organization was represented by persistent sub-groupings of the biomass into domes of density over particular ranges of body size, irrespective of the different production systems that were sampled. Because of the apparent correspondence of these density domes to the ranges of body sizes of the major ecological functional groups—bacteria, phytoplankton, zooplankton, benthic organisms, and fish—this patterning implied that the spectral structures so defined must be linked with general features of the control of energy flow, hence related to the food chain linkages that make up the networks of trophic relationships.

The fact that these density domes within the spectrum appear over similar body-size ranges in different environments supporting differ-ent species confirms that they are independent of the particular species compositions of communities. That is, ecosystems appear to consist of functional, trophic ecosystem groupings by types of species. A similar development of a stable trophic flow of energy mediated by different species in various stages of repopulation of islands devastated by hurri-cane damage was described by Heatwole and Levins (1973), who in-ferred a conservation of trophic function in ecosystems, mediated by species that found opportunities to flourish at different stages in succes-sional community redevelopment. Since then the functional roles of various species groups have been studied in detail in relation to envi-ronmental changes (Frost et al. 1995). It appears that conservation of function mediated through complementary species changes must be reflected in the stable and characteristic profiles of biomass body-size spectra.

In keeping with a terminology suggested initially by Levine (1980) to describe those commonly encountered trophic groups of organisms of somewhat similar type that are subject to much the same predators and feed on similar types of organisms as prey, we shall identify the suc-cessive domes in ecosystems by the term "trophic positions" as a way of recognizing their prominent place in the biomass spectral structure, without necessarily an accompanying knowledge of the specific linkages of which they may be parts. In this chapter we examine the extent to which this internal patterning of positions in the spectrum might be taken to reflect the ecological result of a very general dynamics underly-ing predator-prey interactions.

This possibility had been anticipated by Dickie (1972), who, at the time that Sheldon, Prakash, and Sutcliffe (1972) first described their observations, had been intrigued by the potential for general prediction that their findings afforded. He pointed out, however, that predicting the production in an ecosystem requires an explicit formulation and testing of ecological *population* relations within the food chains. In an exploratory formulation of the elements of predator-prey systems that must be involved, he discussed several features that need to be taken into account. Their importance has been confirmed by later observation, and their implications have gradually been taken up in renewed theoretical development in relation to the spectrum.

## The Relation of Production to Body Size

Dickie (1972) noted that the first and most obvious feature of the organization of productivity relations that needed fuller examination was the practical as well as theoretical difficulty of defining and counting distinct trophic levels in relation to any particular species or predator-prey pair. Identification of trophic level was an essential step called for in production analyses according to the seminal formulation of ecosystem tropho-dynamics by Lindeman (1942). Through the subsequent decades, it had gradually become apparent that this simple, heuristic concept was virtually impossible to apply because of the many functional roles played by a given species at different times in the complex and seasonally variable natural production systems common in aquatic environments. A classic example had been detailed by Hardy (1924), who illustrated the many ecological roles played by the abundant herring (*Clupea harengus*) during their development in marine communities. These realizations led to increasing use of the concept of food webs to replace the apparently oversimplified, but appealing, notion of food chains.

Difficulties in determining the number of trophic steps involved in the food webs leading to fisheries production had already been shown to be a major obstacle in objective explanations and predictions of yield from the Peru Current ecosystem. In the 1960s this system was the most important fish producer in the world, and one that attracted scientific attention because of the apparent simplicity of its particular food web (Paulick 1971). The complexities and uncertainties of interpretation that appeared led Ricker (1969) to propose that the food chain length estimates required for calculating the production of exploited fish species from measured phytoplankton production might best be approximated in the webs by adopting fractional trophic levels.

This line of thinking has been considerably developed in recent years in mathematical analyses of food webs of the sort outlined by Kay, Graham, and Ulanowicz (1989). One of the main problems, from the point of view of comparative ecology, is to arrive at a measure of the dynamics that is independent of the particular structure necessarily imposed on the system by the investigator. The prospect that utilization of the objective measurement of body size might obviate some of these subjective logical, observational, or computational difficulties gives the concept of a size-based spectral representation of aquatic ecosystems a considerably enhanced scientific as well as practical significance.

A second feature noted by Dickie (1972) emerged from the observation that, in calculating production of groups of predators from their prey, the food intake and growth efficiency of the predator were as responsive to the relative body size of the prey as they were to differences in prey density. The results of this fact interacted with apparent ambiguities in defining what was meant by the trophic level of a particular prey species to confuse the formulation for calculating production. The important role assigned to body size by Sheldon and associates as the defining, abscissal feature of the spectrum made it a viable alternative that might also help overcome such ambiguities in fisheries prediction.

Third, Dickie (1972) pointed out that while the estimated ecological efficiency of energy transfer within food chains seemed to fall within quite narrow general limits, this overall apparent stability was based on several underlying parameters that are themselves strongly body-size dependent. That is, energy transfer in food chains can be considered as a function of three size-related parameters (see figure 4.1, a schematic based on Dickie, Kerr, and Boudreau 1987): the rate of predation by the predator $(F)$, the prey growth efficiency $(K)$, and the specific rate of production of the prey $(P/B)$, provided, of course, that estimates of the specific rate of production are population measures, correctly based on changes in the measured densities of cohorts of organisms in their natural environments, rather than on physiological observations of the growth rates of individual organisms.[2]

Following the reasoning of Waters (1969) and Mann (1969), respecting average values for lifetime specific production in animal cohorts in relation to their average body size, and combining it with Bonner's (1965) calculation of the relationship between body size and generation time in animal species, Dickie (1972) suggested that one might expect to find a relation between the annual specific rate of production and body size in fisheries production systems that was not a simple function of the allometric relation between specific metabolism and body size, generally accepted as being about $-0.25$. In a simple model utilizing data on the growth of cohorts of fish under natural circumstances, he calculated that a value for the allometric exponent of about $-0.37$ fitted the data. This value is remarkably close to what was later verified by Banse and Mosher (1980) in their analysis of extensive data on invertebrates, to which they were able to add some additional data for fish and homeotherms.

The fact that ecological energy transfer efficiency is strongly dependent on the relative body size of predator and prey had been recognized by Paloheimo and Dickie (1965, 1966a,b) as a result of their analyses of the growth of fishes under various experimental feeding regimes. The ecological significance of this fact to production in nature was made apparent by Kerr and Martin (1970) and more specifically by Kerr (1971a,b,c), who examined the effects of food type and food chain length on yield and natural production on whole lake ecosystems. These analyses showed that when a predator shifts its feeding from a small to a larger body-sized prey, such as from zooplankton to small fish, its efficiency of capture may increase so much that it completely compensates for decreases in transfer efficiency expected from the additional step in the food chain (i.e., the predator is feeding on fish prey that are themselves predators on zooplankton). From such observations it became clear that trophic level is not the primary determinant of the efficiency of overall food energy conversions in the organization of food chains.

We deal with the organizational analysis of food chain systems in more detail in chapters 5 and 6, when we have much more detailed information available. For present purposes we need only note that the most important implications of these food chain studies were utilized in Kerr's (1974) formulation of the overall slope of the spectrum, where he expressed energy transfers in terms of differences in the metabolic rates of the different body sizes of predators and their prey. In this formulation Kerr made the first decisive step away from the restriction of production analysis to the calculation of trophic levels, replacing Lindeman's transfer coefficients in relation to trophic level by measures of the predator-prey body-size ratio and attendant estimates of efficiency and population specific growth in relation to body size. His results amply justified the generally low spectral slope that had been found empirically by Sheldon, Prakash, and Sutcliffe (1972). Through this recognition of the relation of the production in ecosystems to body size of the component organisms, the stage was set for development of a model of general ecosystem dynamics utilizing the organization implicit in the body-size distribution of biomass density in the spectrum, rather than depending on the compartmental analyses of food webs, which are inevitably liable to the subjective judgments of the investigator.

## The Related Population Density Effects

There remained in the literature, however, considerable confusion about the role of density in determining specific growth rate, hence of calculating the associated parameters of population production implied by the spectrum. For example, following the study by Paloheimo and Dickie (1966b) of the effects of rate of feeding on growth efficiency of groups of fishes in various experimental feeding situations, Calow (1977) had pointed out that certain of the results did not accord with what would be deduced from changes in physiological efficiency by individual organisms during their life history. As he put it, organisms are not "passive valves" that can be modeled by assuming a constant growth efficiency throughout their lives. Paloheimo and Dickie (1965) had already pointed out that this change in efficiency among growing *individuals* is not denied; however, the experimental results confirm that *population* growth efficiency is independent of body size. The apparent correlation between individual and population effects in certain data is a population phenomenon that needs to be interpreted in terms of what can be called the "main sequence" of events to which growing cohorts of a species are exposed, given constant conditions of food availability.

The effects become clearer from experiments where there is constant food availability—for example, in situations where fish are allowed to feed to satiation once a day, so that it is the predator rather than the experimenter that determines the predator's food intake. In such situations, as a cohort grows in body size its rate of food consumption increases, as long as the number of individuals in the cohort remains relatively constant (i.e., low mortality rate). Under these circumstances there is an inevitable inverse correlation between rate of feeding and growth efficiency (defined as $S/pC$, eq. 6.2), which produces an apparent negative correlation between efficiency and body size, which is not, however, directly causal. That is, experiments that reflect natural feeding conditions show that growth efficiency changes only in response to the change in "availability." It is availability—reflected in, for example, changes in the frequency of exposure to food where the particle size or the distribution of the prey changes—that determines the rate of food intake, hence population growth efficiency. In nature, food intake by groups will appear to be related to changes in body size because of the concomitant changes in density and body sizes within the schools of fish that are feeding together. The effects may be further complicated by changes in density distributions and body sizes of the prey. Because of the inevitable and possibly complex changes in distribution and density within the ecosystem, Paloheimo and Dickie (1966b) concluded that population models of the feeding and growth of fishes need to be explicitly defined in ecological rather than physiological terms. That is, in the whole ecological setting, the production that arises through the physiological characteristics of individual organisms will be seen to be continually modified in its ecological expression in the biomass through concomitant changes in the densities of the various particle sizes of organisms in the biological assemblages.

Questions of interpretation arising from confusion between individual and population interactions in natural populations were clarified by the careful work of Humphreys (1979, 1981). He analyzed net production and respiration in some 235 field studies in terms of the densities of organisms per unit area. While Calow had emphasized the life history changes in physiology of individual organisms, Humphreys showed that in populations the net growth efficiency of organisms of similar type per unit area is a constant, independent of body size, trophic level, food quality, or the magnitude of the production ($P$) or respiration ($T$) of animal biomass per unit area. That is, Humphreys (1979) established that the relation between production and respiration per unit area of habitat may be described by a power function according to the equation

$$P = cT^d \tag{3.1a}$$

or in logarithmic terms as

$$\log P = \log c + d \log T, \tag{3.1b}$$

where the fitted slope d of the straight-line relationship takes a value of +1.0 within the various ecological functional subgroups. A summary of his results, which he defined in relation to groups consisting of large noninsect invertebrates, fishes, birds, mammals, and so on, is shown in figure 3.1. The near parallelism of the within-group +1 slopes, but different values of the intercept c among the groups, directly illustrates the ecological differences in average production efficiency among them.

Calow (1977) had pointed out that the physiologically determined net growth efficiency of individual poikilotherms is much higher than that of individual homeotherms, and there is evidence for differences between feeding types, with active foragers being less efficient than "sit-and-wait" predators. Humphreys (1979) showed that when data for these types of organisms are considered in terms of their biomass densities (total body weight per unit area), the differences between poikilotherms and homeotherms become less evident and differences between types of species tend to grade into a continuum in which terrestrial vertebrates have lowest efficiency, fishes and social insects are intermediate, and nonsocial insects and invertebrates have highest efficiency. Within functional groupings, populations of carnivores are more efficient converters than populations of herbivores. However, both classes of feeders are exceeded in population growth efficiency by detritivores, which on the basis of individual physiologies had been previously thought to be the least efficient converters of all.

This important distinction between individual and population conversions of energy in predator-prey situations has not always been made clear in analyses of food chains. The review by Humphreys, and the contrasts between his results and those of Calow, help to clarify the population density status of the phenomena that were described in the orig-

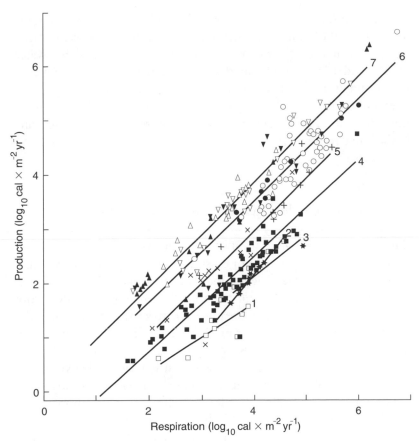

FIGURE 3.1 The relationship between respiration and production (both as $\log_{10}$cal $\times$ m$^{-2}$yr$^{-1}$) in natural populations of animals. Note that these are denoted by $T$ and $P$ respectively in equations 3.1 et seq. The lines are numbered 1 = insectivores, 2 = small mammal communities, 3 = birds, 4 = other mammals, 5 = fish and social insects, 6 = noninsect invertebrates, 7 = nonsocial insects. (Redrawn from Humphreys 1979.)

inal biomass spectrum observations of Sheldon, Prakash, and Sutcliffe (1972) and were being elaborated in subsequent studies. Failure to appreciate the distinction between the energetics of isolated individuals and the energy balance achieved through density adjustments in naturally occurring populations obscures the interpretation of both calculations and reasoning.

The conclusion that the energetics underlying analyses of trophic transfers within the biomass spectrum has to be clearly based on transfers between the biomass densities of the various functional classes of organisms is in keeping with the need for care recognized by Waters (1969) and Mann (1969) in the calculation of specific production (*P/B*), and utilized by Dickie's (1972) analyses. Failure to recognize the effects of changes in density patterns as a necessary ecological unit of measurement underlying biomass density spectra helps to explain some of the initial resistance to accepting the very concept of a biomass body-size spectrum.

## The Two Scales of Control in Ecosystem Production

Banse and Mosher (1980), stimulated by Dickie (1972), drew attention to the fact that interpretation of ecosystem dynamics requires recognizing two different time scales in the control of biological transformations. They compiled data on the annual specific rate of production (*P/B*) of various groupings of organisms by species, and recognized that energy transformations within and between individual organisms will be reflected in the data. That is, energy transactions in individuals must be limited by the physiologically determined relations between body size and metabolic rate; thus energy transfers mediated through individual organisms will reflect the time scale of physiological growth rates. On the other hand, Banse and Mosher showed that within the invertebrate population group, for which there was most data, the rate of change in the annual specific production in relation to lifetime average body size is significantly higher than would be expected from physiological, metabolic considerations alone.

To deal directly with productivity in systems, Banse and Mosher (1980), instead of using biomass density, formulated the spectral relations explicitly in terms of specific production in an allometric power function:

$$P/B_s = aB_s^b, \tag{3.2}$$

where the annual rate of specific production ($P/B_s$) was an average value determined for species *s*, the body mass ($B_s$) was an average for the species at the age of sexual maturity, and "a" and "b" were constants fitted to the calculated data points (one point per species). The constant "b" is, of course, an allometric coefficient relating the measure of lifetime $P/B$ to the body-size index $B_s$, and is directly related to the widely used specific metabolic coefficient $\gamma$, which we will employ in chapter 4 and later. The physiological body-size rule alone would lead to an expected value of b $= -0.25$. They found a value of b $= -0.37$, as had been anticipated by Dickie (1972). They showed, moreover, that if the specific production data for species within other trophic positions in the food-chain systems, such as the fishes or mammals, were treated similarly, the value of the exponent "b" in equation 3.2 remained relatively constant near the numerically higher value. That is, there was a relatively stronger, constant "within-trophic-position" negative slope of the relation between average body size and specific production (figure 3.2).

Dickie, Kerr, and Boudreau (1987) pointed out that allometric exponents of this higher numerical order are the same as values found to relate the natural density distributions of feeding terrestrial herbivores to their body sizes. An important example is provided by values of the relation between home range and body size reported by Harestad and Bunnell (1979), which were interpreted by Damuth (1981) to show how they reflect varying degrees of "overlap" in the home ranges of different body sizes of animals feeding together. Thus, the appearance of the higher "within-group" slope, describing production changes with body size in organisms within trophic positions, seems to reflect a general adjustment in density of predators to prey in the trophic linkages of both aquatic and terrestrial ecosystems.

As expected, the values of the intercept, "a," of equation 3.2, denoting the average levels by species of $P/B_s$ *within* different trophic positions, are different from one trophic position to another. These displacements are clearly parallel to the displacements in production efficiency among functional groups that had been calculated independently, using different criteria, by Humphreys (1979) and shown in figure 3.1. The within-group intercepts are, of course, quite different from the integral value that can be calculated *between* (overall) trophic positions.

Banse and Mosher did not calculate an integral slope value for specific production over all the trophic positions for which they had data, but a value of b $= -0.18$, in general accord (cf. Griffiths 1998) with expectations of the empirically determined metabolism-body-size rule, was calculated by Dickie, Kerr, and Boudreau (1987) using their data,

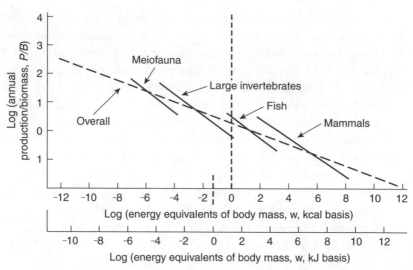

FIGURE 3.2    The relationship between annual specific production and body size within various animal groups, adapted from Banse and Mosher (1980). The mammal and large invertebrate lines are approximately as given in their paper. The fish line is recalculated and redrawn using additional data points deduced from Humphreys (1979). The meiofauna line is based on Schwinghamer et al. (1986). The overall slope (−0.18) is calculated from the data of Banse and Mosher (1980), using their figure 1 and for 5°–20°C. To facilitate comparisons, a scale based on joules is added to the $\log_{10}$ kilocalorie scale used in the original publications. (Redrawn from Dickie et al. 1987.)

supplemented by additional points that were available from the work of Humphreys. Thus Banse and Mosher (figure 3.2) established that there are two time scales of interaction controlling the production processes. These two processes are directly indexed by the two slopes of specific production: one over all trophic positions and the other within trophic positions.

Banse and Mosher (1980), following Dickie's formulation of the relation between *P/B* and ecological efficiency, suggested that the numerically larger, within-trophic-position value of the allometric coefficient of specific production with body size be recognized as an ecological rate scaling—that is, a scaling that reflects the time scales of change in animal aggregation densities. The data that they employed on averages for species did not permit them to perceive events at the scale within the species at which the mechanisms must work, although they made the reasonable inference that elements of mortality are likely to be involved in establishing characteristic density differences among the species that make up a particular ecological functional position. It is clear from their analyses, however, that the anticipated physiological scaling that would reflect the metabolic rate of individual organisms in relation to their body size must be present in the general trend of specific production and biomass over all trophic positions. The stronger ecological scaling that appeared from the plots within trophic positions must reflect the additional adjustments of density that must be made between individuals within a given trophic position.

The similarity between these results for aquatic production and those deduced by Damuth (1981) in relation to overlaps in home range in grazing mammals attests to the generality of ecological population factors that must be added to the physiological limits of individual organism metabolism to support the continuity of natural food chains. That is, adjustments in density are necessary for there to be a balance in trophic transfers between the food requirements of the predatory biomass density and the rate of supply afforded by the prey density.

## Generalizing the Relation of Specific Production to Body Size

Banse and Mosher's (1980) conclusions were taken up and generalized by Schwinghamer and associates (1986) in relation to underlying mechanisms. Their deductions were based on the analyses of benthic communities that Schwinghamer (1981b, 1986) had already provided in relation to the biomass spectrum, supplemented by the observations and interpretations of pelagic ecosystems by Sprules, Casselman, and Shuter (1983). One of their primary interests in the spectrum in relation to benthic production studies was that it should enable them to carry out a functional analysis of their samples without the technically demanding and expensive sorting of the samples into individual species prior to composing body-size density distributions. However, there was the problem presented by Banse and Mosher's (1980) analysis. They had adhered to Dickie's method of calculating average cohort or lifetime $P/B$ for a species in relation to a standardized estimate of its average body size, taken as the body size at which the species reaches sexual maturity. The original objective of such averaging was to eliminate possible effects of age distribution or special species growth patterns on the estimates of $P/B$ among species, but the result was that a calculation of the rate of change in $P/B$ with body size, according to equation 3.2, could utilize only a single point for each species. Such a limitation was clearly wasteful of the information on population structure that was afforded by their data collections and might be utilized in biomass spectra.

Accepting that the allometric slopes estimated by Banse and Mosher (1980) pointed to the dominant general effects of body size within a trophic position, independent of the secondary influences of growth and maturity patterns with age, Schwinghamer and associates (1986) proposed to relax Banse and Mosher's restriction to the "lifetime" $P/B$ of a species and instead developed, in keeping with the plots of Sheldon, Prakash, and Sutcliffe (1972), a general relationship between body size and average annual production and respiration for the biomass density of all benthic organisms that occurred within given size classes. For this purpose they used monthly observational data from their own fieldwork on the mean biomass density of bacteria, microalgae, micro- and meiofauna, and macrofauna to determine an "observed" production. The observations of biomass density could also be used to calculate production and respiration using estimates of the allometric $P/B$ and $P/T$ derived from the literature on the many individual benthic species. By this means they obtained for comparison two sets of estimates, one directly observed and one calculated, as measures of benthic population $P/B$ by size classes for the entire community.

Schwinghamer and associates (1986) then studied the value of their data sets as means of estimating production by comparing the two alternatives: conventional production estimates derived from the sum of productions by body size for each species, with an estimate of production by body-size categories over all species. The results demonstrated the reliability of the overall biomass density method for studying production dynamics without resorting to the initial species sort. There was good agreement between the observed and calculated estimates of $P/B$ by size class for all but the bacterial community, where no good estimates were available of the proportion of the population that was metabolically active at any given time (cf. Gaedke 1993). For the larger organisms, the results enabled Schwinghamer and associates to establish the similarity between their findings and those of Banse and Mosher (1980).

With their new methodology, Schwinghamer and associates (1986) also had available for analysis data on $P/B$ ratios for the several distinct body-size groupings of benthic organisms that had been found by

Schwinghamer (1981b, 1986) in his original biomass density investigations, where he had separated the sampled biomass into three groups:

1. microflora or bacteria,
2. interstitial microfauna, and
3. macrofauna.

The groups had been partitioned on the basis of distinct troughs that appeared as low-density distributions of organisms in the size classes between the three biomass domes. Schwinghamer (1981b) had speculated that these "gaps" represent breaks in the composition of the environment where organisms cannot be supported.

While insufficient data were available on the bacterial community, Schwinghamer and associates (1986) were nevertheless able to establish an integral $P/B$ allometric slope based on the entire community in the benthic biomass spectrum by utilizing published averages to supplement their observations. Using the basic form of the allometric equation in equation 3.2, but dropping the subscripted designation $s$, for species, they found a value of the integral exponent b $= -0.21$, close to the value of $-0.18$ calculated by Dickie, Kerr, and Boudreau (1987) from the augmented Banse and Mosher data series.

In keeping with the earlier findings, Schwinghamer and associates (1986) also analyzed the data for $P/B$ and body size within each of the two larger size groupings representing the meiofauna and macrofauna. The calculated within-group slopes were steeper than the integral slope, although about the same for the two positions. They estimated a common within-position average value of b $= -0.33$, based on data with considerable variability. Their observations thus confirmed the generality of the conclusions of Banse and Mosher (1980) that, using subgroupings of organisms analogous to those identified by Humphreys (1979), the second, ecological rate-scaling appeared in the specific rate of production.

## Biological Significance of the Scaling Effects

While recognizing that the principal organizing force in the patterning that characterized the biomass system must be the energy flow itself,

Schwinghamer (1986) deduced that secondary environmental factors may also be involved. Thus, he concluded that the benthic community is partitioned into functional groupings by body size partly on the basis of trophic relations among the organism body sizes, and partly on the relation of these body sizes to the spatial structure of their physical environment, in this case the particle sizes of the substrate. This appreciation of the effects of the external structure of the physical, benthic system was in accord with speculations on a broader scale by both Schwinghamer (1986) and Sprules, Casselman, and Shuter (1983): namely, that on the basis of apparent relations between the size compositions of the benthic and pelagic elements in the food chain systems, a further separation of biomass spectra into types might result from organism behavioral adaptations in the environment. For example, a primary morphological distinction can often be made between benthic or pelagic means of locomotion and feeding. In combination their data showed signs that there may be a complementary relationship between the biomass spectra of contiguous benthic and pelagic systems.

In chapter 2 we have already pointed out that differences in the positions of the integral spectra seem to be related to environmental differences in nutrient availability. Study of the internal structure of the spectrum now suggests that further distinctions can be made among the underlying causes. The different scales in the dynamics that are reflected in the different levels of organization of the $P/B$ relationships, in particular, indicate two classes of mechanisms operating in the system. That is, the array of stable domes of biomass density along the integral spectral slope indicates that the environmental effect must be mediated by the specifically physiological common factors that control the general metabolic rate of all organisms. By contrast, the characteristics of density change that are reflected by the shapes and spacing of biomass domes, and are reflected in the higher allometric coefficients for $P/B$, can be seen as reflecting ecological factors governing energy acquisition, transfer, and distribution through the biomass densities of the organized ecosystem. Such ecological factors might be expected to be particularly important in relation to ecosystem management.

This recognition of a twofold scaling in the processes underlying production has received interesting confirmation in recent theoretical speculations about the nature of factors that underlie characteristic allometric values relating organism density and body size. West, Brown, and Enquist (1997), Enquist, Brown, and West (1998), and Damuth (1998) have all pointed out that exponents related to the usual value of −0.25 found in physiological studies among both plants and animals suggest the dependence of energy utilization on a morphological or anatomical limitation on energy distribution within the bodies of living creatures. On the other hand, allometric exponents of about −0.33 that we find among ecological groupings appear logically to derive from some form of competition for resources in space. This is clearly parallel to our conclusion that physiological factors are represented by the relation between density and body size over all individuals in the spectrum, while resource limitations that require predator-prey density adjustments are reflected in the "within-trophic-position" allometric slopes. We can only conclude that such a difference in slopes among different degrees of aggregation of production and biomass observations is an essential step in applying knowledge of production processes to the explanations and predictions required for the management of ecosystems.

With their opening of the definition of $P/B$ to the biomass density of all organisms of a particular size class in the habitat, Schwinghamer and associates (1986) had also enabled trophic energy analysis to complete the break with species-specific trophic "level" designations begun by Kerr (1974). They were fully aware of the implications of their findings for the interpretation of trophic energy flows, including the power for formulating new hypotheses about the mechanisms of the control of energy flow that may be involved (Schwinghamer 1983). That is, their results confirmed the generality of the deductions of Kerr and his associates that an analysis based on body size permitted a functional description of the underlying energy transfers within the overall spectrum in relation to the factors that act as mechanisms to control the organization of that functioning.

This synthesis of observations of spectral properties of specific production at the two scales also finally confirms the expectation that the production spectrum is the dynamic base on which the theory of the biomass density spectrum itself might be redeveloped. It remains in this chapter for us to confirm these deductions more generally from the subsequent literature and to reformulate the description of the processes of energy transfer in a manner that better reflects this clearer understanding of the organization of energy flow through the trophic networks.

## The New Empirical Evidence

Following these background developments, a number of papers have examined different manifestations of the biomass spectrum. Some provide confirmation of what has been concluded above. Others point to differences from expectations. We review them here to find what is required for elaborating a more comprehensive theory.

### The Supportive Evidence

Following the publications of Schwinghamer (1981b, 1983), a number of papers attested to the generality of a distinctive, stable, internal pattern or spectral assemblage of particle sizes in aquatic sampling. Among the first were the observations of a variety of benthic systems by Warwick (1984). He examined the contribution of individual species to the body-size composition of the biomass in various localities. While his results attested to the similarity and stability of spectra from different environments, his information on the roles of the different species led him to question the importance ascribed by Schwinghamer (1981b, 1986) to the particle size of the substrate as an organizing mechanism. He concluded that in the organization of the ecosystem of any given area one also needs to recognize the necessarily strong historical interactions of life history strategies that influence the co-occurrence of species in a given community. Both the occurrence and relative densities of any species appear to be as much related to other organisms in the community as they are to the opportunities afforded by variations in substrate composition.

New observations were also made by Gerlach, Hahn, and Schrage (1985) on the structure of benthic assemblages in the Helgoland Bight. They confirmed that the internal body-size structure of the benthic communities was similar to that observed by Schwinghamer and Warwick, with what they regarded as minor differences in the precise sizes at which the gaps in body size occurred between the meiofaunal and neighboring biomass domes. That is, they were able to confirm strong parallel developments of benthic faunal functional organization in quite different areas, based on different species.

At about the same time, Rodríguez and Mullin (1986) published data for a wider range of particle sizes sampled in the pelagic North

Pacific Central Gyre. While their presentation was particularly concerned with examining the robustness of the concept of an integral slope like that predicted by Platt and Denman (1978), in their data there was no mistaking a distinct internal patterning or partitioning of the biomass density within the sampled body-size range, much like that found by Sprules, Casselman, and Shuter (1983). Witek (1986) and Witek and Krajewska-Soltys (1989) also described the internal structural organization for the spectrum of pelagic systems, noting its apparent dependence on various time scales of variability. In addition, Rodríguez, Echevarría, and Jimenez-Gomez (1990) found that in an isolated, oligotrophic, high mountain lake a biomass spectrum had developed that had a clearly defined, internal structural organization parallel to that found in other areas. Similar internal structure was confirmed for the macroinvertebrate community of a lake by Hanson, Prepas, and Mackay (1989) and for the total range of living organisms (albeit mostly small planktonic organisms), including bacteria, in a small lake by Gasol, Guerrero, and Pedrós-Alió (1991).

Boudreau and Dickie (1989) extended the empirical size range of applicability of these spectral constructions, which up to that point had been largely based on data relating only to the smaller body-size ranges of the plankton, to include the fish. In particular, they added to the partial spectra for the Scotian Shelf cohorts of the larger-size fish that make up populations of the common, commercially exploited species. These analyses, supplemented by Boudreau, Dickie, and Kerr (1991), Sprules and associates (1991), and by Boudreau and Dickie (1992), confirmed, on the basis of the empirical information alone, that the domelike structures or positions in the spectrum appeared with regularity in various whole ecosystems in different parts of the world. It seemed reasonable to conclude that the data demonstrated the organization of ecosystems in diverse aquatic environments into domes of trophic position, which suggest similar patterns of regularity in the body-size ratios between the successive functional groupings of predators and prey that compose the communities.

## Deviations from the General Case

Development of scientific understanding is often based on the confluence and an eventual balancing of two opposing tendencies in a manner that supports the emergence of new understanding. On the one hand are perceptions of pattern or generality in what may initially have appeared as confused or complex observed variations. The formulation of the biomass body-size spectrum and the initial generalizations about it on the part of Sheldon and associates (1972, 1973) represented an impulse of this sort. Such observations open the system to interpretations of possible underlying mechanisms, which are a necessary base from which to examine the possible general applicability of the emerging hypotheses. However, generalization over variations on the part of one observer inevitably gives rise to a new view of the significance of variations by some other observer; variations in structures, considered by some to represent a primary pattern, lead to questions about the validity of the original generality and a search for both confirmatory examples or formulations of explanations that would allow tests of the hypotheses. In this way, the variability through which pattern was originally discerned can secondarily become a source of new information, reflecting newly detected effects that alter or obscure the simple manifestation of primary causes. These countervailing tendencies can be discerned in the inquiry about the internal structuring of the biomass spectrum that followed the Sheldon description. The new observations have revealed a wider range of questions than were anticipated in the initial representations and explanations.

## Tests of the Continuous Flow Model

It needs to be appreciated that some of the integral approaches to the spectrum mentioned in chapter 2 implied that only the primary or overall properties of the spectrum have ecological significance. That is, the so-called internal structure of a spectrum was regarded as essentially a form of "noise" resulting from temporary perturbations in systems observed at different times. This appears as the main point in the two papers by Silvert and Platt (1978, 1980), who sought to provide mathematically precise explanations for the internal structure of a spectrum in a form that would be compatible with the continuous energy-flow model of the spectrum proposed by Platt and Denman (1977, 1978). An additional reason for their papers was their recognition of a general need to translate from the essentially static, time-independent models of the biomass spectrum to a time-dependent version that might provide information on production directly from the biomass plots. This worthy objective has also been implicit in a number of subsequent presentations (Vidondo et al. 1997; Han and Straškraba, personal communication, June 1998), among which the alternative suggested by Gin, Guo, and Cheong (1998) that utilizes the two allometric scalings derived from these *P/B* studies offers some encouragement.

As we have pointed out earlier, the Silvert and Platt model took the view that respiration and growth of individuals are the primary mechanisms of energy flux within and between sequential size groups defined in the aquatic ecosystem. As a consequence, one might expect occasional biomass density peaks to appear in an otherwise smooth biomass density spectrum as a result of the commonly observed fluctuations in abundance of particular species, but such peaks would be expected either to propagate uniformly up the spectrum or to gradually diminish over time. Considering that there are other influences on energy flow in the spectrum, such as the effects of a predator on possibly variable sizes of food, and the backward flow of reproductive materials in the spectrum from adults to young, Silvert and Platt hypothesized a smoothing tendency that would damp the successive, but temporary, expressions of any such peak. Their elaboration of the model suggested that, with time averaging, bumps that appeared initially in a distribution would tend to be smoothed out, allowing the overall primary, linear structure that depends on the allometric metabolically regulated energy flows to be more clearly discerned.

This Silvert and Platt elaboration of the Platt-Denman model, evoking a time-averaged energy and material flow as the basis for the spectrum, was specific enough to allow testing, a fact that was discerned in the first instance by Vezina (1986) who undertook test experiments with planktonic organisms. As subject matter he chose to study the rate of radioactive phosphorus transfer among size groups in the bacterial and microplanktonic assemblages, the latter including both unicellular and multicellular organisms. Experimental measurements were made of the appearance of the labeled phosphate in the population elements reared in artificial, cylindrical enclosures floated in an oligotrophic lake. Plankton samples taken during the course of the experiment were divided into 11 $\log_2$ size classes, using screens graduated between 0.4 $\mu$ and 1600 $\mu$. Few zooplankton larger than 1600 $\mu$ appeared in the containers during the short periods of observation.

The data were compiled according to two models of uptake, excretion, and predation loss: one an ordered-flow model that was considered analogous to the Platt and Denman model, one an unordered model that allowed energy to flow discontinuously to various compartments. The results showed that the ordered model led in time to increased deviations of results from expectations. By contrast, the model that allowed energy flow to appear in various compartments gave more satisfactory fits to the phosphorus transfer data. The implied analogy of the unordered model with transfers to either seston or to discontinuous predator compartments suggested a need for improved models that would allow for short-term energy flow in less sequential size-compartment terms than the Platt-Denman model called for.

A result analogous to that of Vezina was obtained by Rodríguez and associates (1987) over longer time scales. They examined the detailed dynamics of a plankton bloom during late winter in Mediterranean coastal waters. Their data showed that particular domes in the size distributions increased abruptly in amplitude during the initial bloom, then equally abruptly subsided to be replaced by other domes, mostly at larger body sizes. That is, the domes did not show progressive changes in position, as had been anticipated by Silvert and Platt. Rodríguez and associates deduced the need to recognize compartments, including detrital compartments, to account for the apparent "bottlenecks" in the flow of biomass or energy up the basal parts of food chains. Gasol, Guerrero, and Pedrós-Alió (1991) noted the same phenomenon in their multiyear seasonal study of a small lake, and Gaedke's (1993) detailed study of seasonal variations in the spectrum in Lake Constance concluded that a discrete-step model fitted the observations better than a continuous flow model. These seasonal descriptions and interpretations of compartmentalization of energy flow confirmed the original descriptions of plankton blooms by Parsons (1969), who first used the biomass spectrum methodology later applied by Sheldon, Prakash, and Sutcliffe (1972) to account for the lack of progressive sequences in particle size distributions in a plankton bloom.

The results of Parsons, Vezina, Rodríguez and associates, Gasol and associates, and Gaedke in combination suggest that the simple continuous flow model fails to capture the main features of production dynamics on a time scale sufficiently sensitive to account for the dominant internal structural details of the biomass density spectrum. This is perhaps to be expected among the smaller organisms with turnover rates that are high relative to the usual, practical time spacings of sampling regimes. In conjunction with the speculations of Sprules, Casselman, and Shuter (1983) and Schwinghamer (1986) about the stable, strong, domelike features they observed in annual averages of samplings, the evidence suggested that the trophic energy flow is channeled into pools of material accumulation, some of which may be detrital; or in the case of interacting benthic and pelagic communities, be reflected in body-size phases of predators and prey, or even in life history stages of organisms that utilize pelagic and benthic phases of the aquatic ecosystem at different times. That is, at the base of the production system, the expectation of domes of biomass density gradually shifting in body size appears to be refuted by the observations, perhaps to be replaced by compartments or domes with significant temporal stability.

It is certainly too early to fully assess the validity of alternative approaches. There is convincing evidence that careful sampling in certain environments can detect significant progressive movements of local biomass peaks. Brander and Hurley (1992) found seasonal evidence for changes in the occurrence of fish larvae of different sizes over a wide area of the Scotian Shelf. Hargrave and associates (1985) also demonstrated for zooplankton that there was progression in the peaks of body size in a small bay where the residence time of the water masses was about 100 days, although in some cases the changes in average size were found to be related to changes in species composition. We can only speculate that the mobility of water masses in relation to feeding opportunities may make the logical generalizations about the continuity of observed systems somewhat suspect in local practice. The apparent early successful use of size-based models that allow for species changes but incorporate the two allometric scalings suggested by the *P/B* relationships may yet become a viable methodology that avoids the problem of time independence without undue mathematical simplification (Gin, Guo, and Cheong 1998)

### Reexamination of the Slope in Relation to Functional Spectra

There are a number of more recent studies of partial spectra that offer additional information on spectral structural properties. We begin with the observations reported by Quiñones (1992) and Quiñones and associates (1994). They compiled both biomass and metabolic-rate spectra from stations in the northwest Atlantic and the western Mediterranean, respectively. In Quiñones' data for 9 stations in oligotrophic regions of the northwest Atlantic, and 10 stations in the highly productive region of Georges Bank, the overall size range from bacteria through mesozooplankton is well fitted by straight lines. However, in parallel with results of the study of Rodríguez and Mullin (1986), there can be no mistaking the internal domelike density distributions, representing three major classes of organisms that would have dominated an unnormalized presentation of the data. In particular, their biomass data show an apparently distinctive bacterioplankton dome in all samples, along with two less clearly separated phytoplankton and zooplankton domes, thus effectively extending the structured size range of organisms included in oceanic pelagic sampling to the bacterioplankton.

An additional important result of Quiñones' biomass study was the apparent uniformity of the overall slopes between the major regions. That is, while the overall biomass densities of oligotrophic and eutrophic regions show the expected large differences in average level, the calculated spectral slopes are almost the same for both oligotrophic and eutrophic regions. This supports the findings of Boudreau and Dickie (1992), described in chapter 2, where a remarkable result of the fitting of spectra from seven ecosystems of differing productivity was the uniformity of slopes, despite the large apparent differences in overall level. Quiñones' studies also showed that the overall unnormalized distributions of biomass densities in all areas were relatively flat, much like those found by Sheldon and associates and in the Boudreau-Dickie study. In normalized terms the slopes were close to, or in most cases slightly less than, $-1.0$, as was found by Sprules and Munawar (1986) for the highly productive regions. The zooplankton range of the body-size distribution of their spectra tended to exhibit a lesser slope than that calculated for the bacterio-microplankton size range.

Equally interesting from the point of view of this book is the fact that these two studies undertaken by Quiñones and his associates represent the first attempts to directly compile a body-size spectrum of the respiration rate of organisms. The results confirm the emerging picture of a spectrum that is based on both physiological and ecological scaling effects. Quiñones (1992) sampled two stations on Georges Bank and four stations in deeper water near it; Quiñones and associates (1994) studied three stations in the western Mediterranean. Bottle samples from various depths were combined, size-fractioned, homogenized, and treated for later direct analysis of their electron transport activity (ETA). According to Packard (1985), the ratio of potential metabolic activity to respiration given by the ETA analysis will be distinctly different for bacteria, phytoplankton, and zooplankton, such that the slope measured in these experiments would likely be a minimal estimate of the actual allometric coefficient of respiration with body size. However, the results are of considerable interest. The slope values for the ETA spectrum measured for the northwest Atlantic averaged $-1.20$, while those from the western Mediterranean averaged close to $-1.34$. Quiñones compared them with slope estimates of about $-1.3$ derived from the literature.

As we have pointed out above (equation 3.1 and figure 3.1), Humphreys (1979) showed that, within certain ecological functional assemblages, population density estimates of respiration and production have the same relation to body size. Therefore, these respiration spectral estimates of Quiñones and his associates may be regarded as the first direct estimates of an index of the production density spectrum for a population in its natural surroundings. While the actual slopes are likely to be too flat to represent final estimates of either respiration or production spectra, they nevertheless provide estimates that are directly comparable to the secondary or ecological scaling of the production spectrum, as it was developed by Banse and Mosher (1980) and Schwinghamer and associates (1986). That is, given adjustments for respiration appropriate for ETA activity estimates, the production spectrum data within the bacterial and phytoplankton body-size ranges appear to reconfirm the relatively steep secondary or ecological allometric slope of the production spectrum of $-0.37$ (normalized $-1.37$), deduced by Dickie (1972) and measured by Banse and Mosher.

The need for a more explicit formulation of the underlying dynamics of the spectrum as a basis for understanding variations in spectral parameters appears in the study by Zhou and Huntley (1997). Their data are restricted to the partial spectra represented by the zooplankton. They adopt the continuous flow model of Platt and Denman as the basis for their formulation of a series of spectra compiled for various subareas of the California Current. Their normalized plots of zooplankton biomass distributions for the individual subareas are all clearly in the form of domes that show strong, regular curvature for the smaller plankton, but irregularity for larger sizes, a fact that they attribute to under-sampling at the upper end of the size range. Their analyses take the form of studies of the rate of change of the biomass or production at each body-mass size class. They interpret this instantaneous slope in terms of a special parameter defined as the ratio of an index of population increase to the weight-specific growth rate. While the biological meaning of their parameters is open to question, the pattern of regularity in overall results led them to calculate an averaged "system" slope for the California Current system, and to look for relations between deviations from it and local current patterns, such as eddies, jets, or meanders. The methodology is too new to allow us to evaluate it. However, the whole approach is significant in the present context because of its implication that details of the patterns within the spectral structure have a quite precisely definable biological meaning. We shall return to considering the value of information at this scale in chapter 5 and 9, after developing a discontinuous, predator-prey model of the spectrum in chapter 4.

## Other Views of the Regularity of Slope and Structure

Further sets of analyses offer additional important information on the nature of the detailed structure of the biomass spectrum. In these cases, moreover, the philosophy that appeared in the valuable critical approach of Vezina (1986) reappears, suggesting that we must reconsider the assumption that the overall slope of the biomass spectrum is a straight line of negative slope. In particular, additional questions about how to best use observations for the formulation and testing of hypotheses of biomass spectrum mechanisms lie behind the analyses of both Ahrens and Peters (1991a,b) and of Cyr and Pace (1993). Their observations and openly critical assessments of generalizations about the nature of the spectrum were confirmed by a number of other investigators, and lead us to look beyond its initial generalized form.

Ahrens and Peters (1991a) assembled data on phytoplankton and zooplankton for a number of lakes, and showed that in most cases the biomass spectrum drawn for them had a higher biomass density of zooplankton than of phytoplankton. This sampling result was consistent with the high zooplankton densities found earlier by Sprules and Munawar (1986) in a number of the more productive water bodies for which they had assembled plankton data. It is also consistent with markedly high, intermediate microplankton domes found by Gasol, Guerrero, and Pedrós-Alió (1991) in their seasonal sampling of a small lake, as well as with some of the data for marine areas examined by Quiñones. Even more dramatic deviations from the simple negative slope of the partial spectrum were shown in the studies of Rojo and Rodríguez (1994) in a 60-week seasonal study of the body-size spectrum of a hypertrophic lake. The resulting partial spectra had positive unnormalized rather than negative slopes (normalized $b < -1.0$), in the case of Rojo and Rodríguez a normalized slope of the order of only about $-0.40$ for the entire seasonal series. Clearly, these populations had distributions of biomass with body size that were not in accord with the normalized negative slopes of the order of $b = -1.2$ that had been deduced from allometric metabolic considerations by Platt and Denman, or even with cases where there was a lesser negative slope of $b = -1.0$ to $-1.05$, which appeared to fit both the empirical results of Sheldon, Prakash, and Sutcliffe (1972) and were the subject of explanation in the later empirical results and theoretical speculations of Kerr (1974), Schwinghamer (1986), Borgmann (1987), Minns and associates (1987), Sprules and associates (1991), and Boudreau and Dickie (1992).

The spectra of Ahrens and Peters, in keeping with the other, largely freshwater examples, were clearly subject to influences that could not be subsumed under the simple allometric effects that predict unnormalized slopes of less than zero. However, there was no indication from either the sampling or from any of the situations examined that the positive slopes of the partial, unnormalized spectra—reflecting biomass domes of the smallest zooplankton body sizes that were both higher and broader than the principal domes at the phytoplanktonic body sizes— in any sense represented abnormal or exceptional ecological situations. In the case of the study of Rojo and Rodríguez (1994), where the lake contained no fish predators, it had even appeared that the slope of the planktonic spectrum might become increasingly positive in slope as it approached its midseason equilibrium condition.

In this connection, it is particularly significant that similar findings of relatively high zooplankton domes were found by Sprules and associates (1991) in their sampling of the whole populations of Lake Michigan where there were fish predators present. In their extensive set of data from the pelagic zone, they showed, however, that while the zooplankton dome was higher than the phytoplankton dome, the overall slope that resulted from the addition of data on the next larger dome, representing the fishes, gave rise to an overall spectrum with a negative slope. This observation, in association with the observations of Boudreau and Dickie (1992) on the spectra for major world fisheries-related ecosystems, in which the zooplankton and benthos dome averages are variable in height, indicate that, at the very least, there are limits to how well an entire spectrum can be reconstructed as a simple straight-line slope from partial sampling. Clearly, the observations of Ahrens and Peters supported the indications of many other studies that the theory of the spectrum required development in relation to the observed densities of various domes if it was to be considered a completely general rather than an oversimplified oddity arising mainly in marine sampling.

In addition to these complications concerning the slopes, the new observations of more complete fish or invertebrate trophic position spectra also made it evident that predator-prey interactions occur within, as well as between, trophic positions. Hence, description of the biomass density as a series of smooth, dominant trophic position domes does not take into account all of the variability that results from the predator-prey trophic interactions that must occur in aquatic ecosystem energy transfers. If predator-prey interactions, such as described in equation 3.2, could be considered as a primary force in establishing the domelike structures of the complete spectrum, it would be reasonable to expect further evidence for the effects of trophic transfers within the major trophic positions. This possibility adds to the need for both more precise empirical description and more comprehensive theoretical explanation of the characteristics of spectra.

The findings of Cyr and Pace (1993), while cast in the form of an implied challenge to the significance of internal spectral features on the organization of ecosystems, offer significant supportive insights into the detailed nature of the spectrum and its apparently broader base. Cyr and Pace (1993) studied the dynamics of zooplankton communities in the same area studied by Peters and his associates. They described samples of what they defined as 90 zooplankton communities distributed in space and time over 28 different lakes and provided detailed empirical information on the seasonal and area compositions of the zooplankton dome. Their results well illustrate the problems that might ensue from a too simplified or facile description and generalization of actual size compositions or their associated dynamics within the zooplankton, particularly if based on inadequate sampling.

They found that the zooplankton communities exhibited considerable variability in the number and heights of subdomes in size compositions, both within communities in time, and between communities and lakes. The overall average of the whole data set yielded a distinctly bimodal size distribution for the zooplankton, but there were also lakes that supported three distinct subdomes. It was obvious that the size distribution is influenced by a wide variety of factors other than, or in addition to, those invoked in interpretations of zooplankton as constituting a single, homogeneous spectral dome or trophic position. Their analyses revealed no general basis for subdivision of the overall zooplankton dome into a series of subdomes; however, the identification of such mechanisms was not a primary objective of their paper.

Their primary interest was to evaluate the effects of irregularities in observed size distributions within the zooplankton trophic position on its total energy flow. In keeping with this objective, they deduced from their own and much more extensive data in the literature that, to a first approximation, the specific rate of grazing by zooplankton with body size followed a power function like that of equation 3.2, with an allometric exponent of about $-0.45$. Given the high variability of their data, it seems unlikely that this value for the exponent describing food intake was different from the $-0.33$ to $-0.40$ found by Banse and Mosher (1980) for specific production of invertebrates. That is, their conclusions enlarge the applicability of previous investigations of community dynamics by showing that the rate of grazing by this most important class of aquatic predators also follows the productivity relations deduced from metabolic and specific production data.

Cyr and Pace carried their calculations an important, additional step, showing that with increasing body size the compensations between population density and grazing rates that take place within the whole zooplankton community result in a more or less constant rate of assimilation over the total range of zooplankton body sizes. That is, within this important functional group or position in the aquatic spectrum, the dynamics of the overall energy exchanges appears to be much the same as is found within other trophic positions. Thus, while demonstrating the naturally occurring irregularities in size compositions from place to place and time to time within the zooplankton dome, Cyr and Pace also demonstrated that the estimates of total assimilation of food and its production in the community were essentially unaffected by this variability. With ample attention to perturbation in the internal structure of the zooplankton dome, their results confirmed an orderliness in the overall energy flow within the zooplankton and between zooplankton and their phytoplankton prey. The persistence of the zooplankton domes in lakes where there were fish also implies that this stability applies in the face of predation on the zooplankton by their predators in higher trophic positions.

The calculations of Cyr and Pace (1993) take on added significance for our overall study of spectral properties when they are juxtaposed with the results of another study from the same area, reported by Ahrens and Peters (1991b). The latter study assembled data on particle size of the smaller planktonic sizes over six orders of magnitude from pico-plankton to net plankton. From the extensive data, they calculated a between-position, hence integral, allometric slope of total metabolism of 0.73. This is equivalent to an integral slope of specific metabolism of $-0.27$. This estimate is numerically high, compared with the estimates of the integral slope of the production spectrum derived by Dickie, Kerr, and Boudreau (1987) from the augmented Banse and Mosher data. However, it may be juxtaposed with the comparably high zoo-plankton slope determined for the same areas by Cyr and Pace's slope of $-0.45$ for production or rate of food uptake within the zooplankton trophic position. While both the integral and within-position estimates from these studies are high, a bias toward higher values is to be expected from least-squares fittings of data with limited body-size ranges (Griffiths 1998). From our point of view, therefore, the main significance of their findings lies in the fact that the difference between the slopes calculated in the two ways is about the same as differences reported earlier between the slopes for the primary and secondary scalings of production density data. We conclude that their results confirm the twofold scaling that we have earlier interpreted with Banse and Mosher as a physiological and an ecological scaling of production processes.

The data for the areas sampled by Peters and his associates thus give rise to interpretations of structure within the biomass body-size spectrum that are directly comparable to those found elsewhere. At the same time, the variability in the internal structure determining these slopes, as well as the evidence for a proliferation of domes, suggests possibly significant additional structure, at least within the zooplankton. Such observations clearly call for elaboration of the simple models of energy transfer implied by the initial empirical observations and elaborated in the preliminary theoretical developments.

## Summary

A persistent secondary patterning of elements within the overall aquatic biomass body-size spectral organization is well established by the evidence. The main aspects of this spectral organization appear as the well-known domes of biomass density corresponding to bacteria, phytoplankton, zooplankton, and so on. It is evident, however, that certain details of the overall structure are not adequately comprehended in the simple model implied by the fitting of biomass density data by size class to straight lines, as was originally suggested by Sheldon, Prakash, and Sutcliffe (1972) and formulated as theory by Platt and Denman (1977, 1978). Subsequent analyses confirm that this structuring is related to an ecological level of causal relationships that is not evident in purely integral data fittings. Study of allometric relations at two different levels of aggregation revealed in specific production ($P/B$) data leads us to conclude that dynamics of change in ecological densities may be very significant for ecosystem management.

The recent evidence also suggests that at least two features revealed by the internal structuring of biomass spectra need to be considered in any elaboration of the underlying theory. In the first place, it appears that departures from straight-line integral spectral slopes may be the rule rather than the exception where there are detailed data to represent the complete biomass spectrum of sizes. In the second place, domes describing particular trophic positions may not be smooth or uniform. There may be secondary peaks within the major categories, and these too may have biological significance that cannot be detected from integral spectral plots.

In the next chapter, we examine the possibilities for understanding these features by considering dynamic mechanisms that constitute a theory of trophic energy flow. Using specifically defined predator-prey mechanisms, we show that the deviations appear to have arisen from oversimplified initial conceptions of the energy flow in the spectrum. In chapter 4 we examine mechanisms that lead to curved integral spectral profiles, and in chapter 5 extend a predator-prey model to enable us to comprehend the complex internal structuring that emerges from the empirical study of biomass density distributions.

## Notes

1. The compendium of Cyr and Peters (1996) seems to have a very limited value as a test of these ideas. While the conclusions drawn are consistent with ours, the high variability in the data, probably reflecting the heterogeneity of sources, methods, and authors as much as the observed material, is not sufficiently accounted for to base generalizations about the spectrum on them.

2. This is not a trivial stipulation. As late as 1998 the editors of the scientific journal *Oikos* published, without comment, and so evidently without awareness of the error, a paper by Roa and Quiñones in which population production was equated with the sum of individual growths. The mistake was subsequently pointed out by van der Meer (1998) and published by *Oikos* in a section entitled "Opinions"!

# CHAPTER 4

ᓭ

# Toward a Mathematical Model

## Biomass Density and Abundance as Scales of Ecosystem Observation

With the foregoing empirical explorations of the structure of both biomass density and production spectra, reflecting at least two classes of underlying mechanisms operating at different time scales, the way is cleared for considering what Dickie, Kerr, and Boudreau (1987) called an "algebra of community dynamics." This algebraic formulation has a twofold purpose. In the first place, it defines relations of various measurable characteristics of spectra to the pattern of ecological interactions within the ecosystem as a whole. In the second place, it brings together a terminology that has been developed in general ecology and biological oceanography separately from the language used in one of its principal subdisciplines: fisheries dynamics and management.

Differences in terminology do not arise by accident. They reflect potential differences in purpose between parallel investigations in subdisciplines. For example, the general field of ecology and its fisheries management branch have both been concerned to relate a common concept of production processes to the various parts of the ecosystem. However, the sampling methods and objectives of the two subdisciplines have been directed to such different scales of observation that a sense of their commonality has been obscured or lost. Reestablishing a common terminology is an integral step in constructing generally satisfactory theoretical spectral models of the whole ecosystem, even while recognizing that special submodels may be needed if we are to identify controls on energy flow for particular purposes, such as management of the mechanisms underlying fisheries.

The differences in terminology and observation scale are symptomatic of differences in perception similar to those that gave rise to confusion about the causes of changes in production through Calow's studies of changes in the physiology of *individual* organisms, in contrast with Humphreys' studies of changes in the biomass density of *population* assemblages. This confusion seems to have been responsible for delaying recognition of the consequences of different scales of operation of the energetics: energy exchanges due to the metabolism of individuals, and mediation in system energy flow due to the density adjustments among individuals making up the populations. It illustrates the need for special care in identifying both the scale of observations and interpreting the aspects of organization they represent.

In relation to compiling data for the spectrum, we draw attention to a parallel confusion that has repeatedly arisen in aquatic studies in attempts to relate primary and secondary production of planktonic organisms to production of the larger, more widely dispersed fishes. Changes in the sampled *density* of populations of organisms are the basic data used to describe production of the masses of smaller, planktonic organisms in biological oceanography. Since the mid-1950s, however, fisheries studies have been preoccupied with changes in the overall *abundance* of "stocks" of particular species. While density and abundance are logically simply related, the practical failures to recognize the ways in which unmeasured local changes in density obscure the evidence for changes in abundance have led to serious problems in both scientific analysis and management. Applying the biomass spectrum to analysis of production processes depends on explicitly recognizing the differences in point of view implied by these different scales of measurement.

## Comparing Data on Plankton and Fish Density

Studies of plankton and the mechanisms underlying their dynamics employ detailed sampling to define the biological characteristics of particular water masses, often without being able to establish either the volume of the total water mass or the total of the biomass in question. In most instances (e.g., Zhou and Huntley 1997) dynamics of population interactions can be satisfactorily determined from changes in the densities or concentrations of the components. It was this kind of thinking that led in the early 1970s to Sheldon's concept of a biomass density spectrum of body sizes. Its development was directly related to the methods of sampling customary in plankton research.

Sampling and analysis in fisheries has been on a quite different scale because the interests of governments, industries, and markets have been primarily in total landings and economic yields of species. The details of technological efficiency and employment related to densities are also important, but interest in them tended to be secondary. The practical application of scientific analyses therefore required total abundance information, leading to the extensive development of a dynamic description of fish population production based on total fish landings and designed to track and interpret changes in the abundance of the individual species, which generally have quite different economic values. From this perspective, measures of density of species were mainly regarded as approximations of changes in abundance. Logically, this must be true for all homogeneously distributed population elements of a given type.

Equating measures of density and abundance in production studies requires, however, that one have technical knowledge of the heterogeneity of the distribution of both predator and prey elements within a defined stock area. The practical consequences of failures to detect deviations from homogeneity have been slow to be appreciated. In the major Pacific halibut fishery, for example, the need for extreme care in averaging the density index of catch-per-unit-effort was recognized early by Thompson, Dunlop, and Bell (1931) and Thompson and Bell (1934), but only with respect to the relative yields of the different fishing gears used. The importance of undetected distribution effects on the interpretation of abundance changes in relation to fishing was not appreciated for a further 40 years, when it was made clear in the analyses of Skud (1975). In the whitefish fisheries of Lake Winnipeg (Hewson 1959, 1960), and the lake trout fishery for Lake Superior (Lawrie and Rahrer 1972), it was found that differential distribution of the fishing effort over the area of fish distribution obscured the detection of declines in abundance. It took special investigations to bring to light the fact that during the development of these fisheries, fishing vessels moved to progressively more distant grounds as local concentrations were exhausted. Thus the routine statistics that were compiled in terms of landings per day fished gave a falsely high and steady "index" of abundance until the fisheries virtually collapsed.

Parallel effects have been observed on more local scales as new technologies have been introduced. For example, modern fish detection equipment permits a detailed discrimination between the densities within small parts of the general area of distribution. This has enabled fishermen to concentrate their efforts on local high densities of fish, maintaining the average level of catch-per-unit-time until general abundance becomes very low. Failures to detect these effects have been sources of serious confusion both in perceiving the needs for fisheries management and in measuring its effectiveness (Calkins 1961; Saville 1979; Hutchings 1996). They also seem to lie at the base of difficulties in compiling biomass spectrum information for terrestrial systems. We deal with some of these problems and their ramifications in more detail in chapter 8.

An analogous difficulty has developed in attempts to explain observed changes in both total catches and economic returns from both hunting and fishing activities. Early theoretical or experimental models designed to study the success of predators in capturing prey, or of fishing boats in capturing fish, were originally formulated for uniform population distributions. Where the boundary of an experiment is a fish tank, or where a model of a population is treated as though composed of homogeneous units within a constant environment, a change in abundance is the same as a change in density. However, this simplicity is soon lost when distributions are not uniformly or randomly distributed. Overlooking this simple fact has obscured the appropriateness of applying certain mathematical or statistical models to the interpretation of observed population phenomena (Gulland 1962; Steele 1978). In actual fisheries, where a change in the abundance of any aggregation of organisms always seems to be accompanied by a change in the total area of its distribution, measures of abundance and density cannot be equated without specific reference to the details of the area differences (Creco and Overholtz 1990; Rose and Leggett 1991; Swain and Sinclair 1994). In interpreting biomass spectra, these heterogeneities can mean that calculations of relative abundance of predators and their prey may not reflect causal energy transmission phenomena at the level of the mechanisms underlying production.

Because of these effects, detailed information on density distributions has gradually been seen as increasingly relevant to the scientific interpretation of fisheries data as well as plankton studies. The requirement has reentered fisheries data collections in the adoption of extensive stratified random surveys. These are recognized as necessary to "correct" general abundance estimates derived from "virtual" population analyses (Doubleday and Rivard 1981), but the resulting density information has been increasingly perceived as useful to all branches of aquatic research.

A simple attempt to overcome specific limitations in comparative fisheries data using relative area information was described in chapter 2 in connection with the Boudreau and Dickie (1992) compilation of an integral version of the biomass spectrum for several of the major world fisheries. In general, however, methods of direct fisheries sampling have been developed that make density information more reliably available and applicable. During the past two decades these have markedly increased properly weighted average density information related to the whole biological spectrum for a number of areas (Hennemuth 1979; Duplisea and Kerr 1995).

Application of the biomass spectrum to the analysis of natural populations, including fish, also depends on the homogeneity of the trophic system for which a spectrum is being compiled. It is clear from the studies of the internal structure of the spectrum that density calculations for large areas may need to take account of significant local differences in community structure, especially the availability of prey to their predators. The phenomenon of heterogeneity in the parameters of the biomass spectrum was investigated in an extensive marine area by de Aracama (1992) and Duplisea and Kerr (1995), who showed how the degree of definition of parameters of the spectrum may change with the size of area over which data for a system are compiled. We discuss the details and their implications further in chapter 9.

What is needed for the theoretical development that follows in this chapter is assurance that these important sampling problems have been appropriately addressed. The theory may then be used to verify the significance of deviations in the structural details of the spectrum from the general case and to judge the need for modifying the theory to take account of the effects of additional but neglected influences. We can anticipate that these neglected influences may appear more commonly at the smaller scales, reflecting differences in the day-to-day behavior of

aggregations of predators and their prey that could not be comprehended by the less discriminating sampling technologies available to
earlier, more generalized studies.

## Algebraic Models of Aquatic Production Parameters

With the assurance that parameters of change in the biomass spectrum
are specified in terms of densities of the population elements, Dickie,
Kerr, and Boudreau (1987) proposed a terminology to clarify relations
between the densities of predators and their prey in relation to the three
factors that Dickie (1972) had identified as the underlying determinants
of energy transfers within food chains: the efficiency of conversion of
the biomass of prey to the biomass of predators ($K$), the rate of capture
of food by predators ($F$), and the specific rate of production of the population elements ($P/B$). These terms are defined in relation to production in the simplified trophic energy flow diagram given in figure 4.1,
adapted by Dickie, Kerr, and Boudreau (1987) from an original by
Petrusewicz and Macfadyen (1970) and modified for presentation here.

The central axis of figure 4.1 consists of a sequence of measurable
elements making up the various steps in uptake and transformation of
food energy from the primary producer biomass ($B_0$) through an herbi-
vore biomass ($B_1$) and a primary carnivore biomass ($B_2$). At each trophic level (n)—sensu Lindeman—in the energy cascade, the diagram
recognizes that the observed biomass is the result of a balance among the
net growth ($G_n$) that derives from the production of the organisms in it
($P_n$), mortalities of the biomass due to nonpredatory causes ($M_n$), and
mortality due to predators at the next higher trophic level ($F_{n+1}$).

In order to allow the explicit description of various ecological relations, the intake of food (catch) by predators at level n + 1 is specified
as $C_{n+1}$, of which only a certain fraction ($p_{n+1}$) is digested and assimilated ($S_{n+1}$), and a further fraction ($m_{n+1}$) is metabolized to result in the
energy used in production by the predator ($P_{n+1}$). With this scheme,
definition of the parameters associated with the predation process can
be readily agreed on. Thus, the predatory uptake from prey level n,
defined as $C_{n+1}$, is the result of the combined predatory actions on
it, defined as the product of the number of predators at level n + 1,
$f_{n+1}$, and their individual capturing efficiency, $q_{n+1}$. The product,

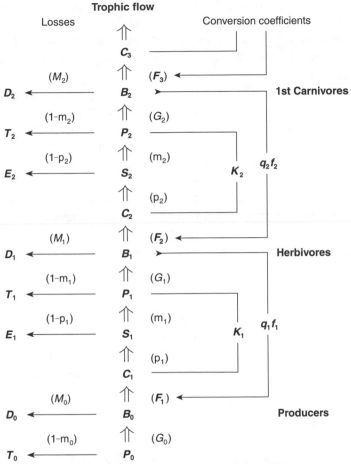

FIGURE 4.1  Energy flow pathways and transfer efficiencies by trophic steps (n) in an idealized food chain. The central trophic flow shows $P_n$ (production), $B_n$ (biomass), $C_n$ (food intake, or catch), $S_n$ (metabolizable energy); on the left are energy losses: $T_n$ (metabolism), $D_n$ (deaths), $E_n$ (eliminated wastes); on the right are conversion coefficients $K_n$ (ratio of produced to consumed energy $= m_n \times p_n$), $F_n$ (predatory mortality $= q_n f_n$); efficiencies are in brackets: $m_n$ (metabolic), $p_n$ (digestive-assimilative), $M_n$ ("natural" mortality), $G_n$ (specific production).

$q_{n+1} f_{n+1} = F_{n+1}$, is thus defined as the rate of predatory mortality exerted on the biomass of the prey at level n: $B_n$. This definition of predatory mortality is directly analogous to the conventional use of the rate of fishing mortality in fisheries to describe the impact of the fishery on the target species.

As can be seen from figure 4.1, the efficiency of production from the uptake of prey, $K_{n+1}$, is a product of the digestion-assimilation efficiency, $p_{n+1}$, and the metabolic efficiency, $m_{n+1}$. The specific rate of production at level n + 1 is measurable as the ratio $P_{n+1}/B_{n+1} = G_{n+1}$.

If we use this schematic to follow the passage of energy through the food chain and designate the biomass density of any predator as $B_n$, and that of its prey as $B_{n-1}$, then the rate of transmission of biomass through successive predator-prey steps within a food chain is measured by the ratio $B_n/B_{n-1}$, which, according to the diagram, is the product of the three successive rate processes between the two biomass elements, thus:

$$B_n/B_{n-1} = K_n F_n B_n/P_n. \tag{4.1}$$

In the detailed discussion of the dynamics of food energy transfer studied in this and succeeding chapters, the two terms, $K_n F_n$, taken together, will be referred to as a "conversion coefficient." The third term, $B_n/P_n$, is the important turnover rate of the predator discussed in chapter 3. That is, turnover rate is the inverse of the specific rate of production investigated in detail by Banse and Mosher (1980) and Schwinghamer and associates (1986) and used as the basic data for the production spectrum of chapter 3. In the algebra of equation 4.1, we are thus able to equate in logical terms the biomass ratio between a predator and its prey with the turnover rate of the predator corrected by a conversion coefficient that describes the efficiency of the energy transfer processes.

Two points of clarification of the production model of the biomass spectrum implied by equation 4.1 are allowed by this explicit algebraic form. In the first place, it can be seen that the ratio of biomass densities can be directly translated into trophic positions in the biomass spectrum if we designate a constant ratio, $R_n = (w_n/w_{n-1})$, between the body size of the predator and the body size of its prey. It is, therefore, clear that the nature and stability of the body-size ratio will be important aspects of the development of a mathematical predator-prey model of the biomass spectrum. This recognition also gives a certain emphasis to the basis for the difference between the predator-prey energy transfer models we are discussing here and the continuous mathematical model of Platt and Denman in which the spectrum was considered to be made up of ratios of the successive body-size intervals arbitrarily chosen by the investigator. That is, we employ the body-size ratio as the basis for converting a generalized mathematical energy-flow system into a model of successive functional steps in the flow up the ladder of prey energy to predator energy within a defined ecosystem.

In the second place, from the point of view of using the spectrum as a tool in comparative ecology, it can be seen by simple manipulation of this algebra that, aside from the important, measurable biomass ratio itself, two other central concepts of ecology, termed the *ecological efficiency* and the *production efficiency*, are measures of energy transfer that are directly analogous to the ratio of biomass densities described in equation 4.1, although taken over different parts of the energy transformation sequences in successive trophic positions in the schematic food chain system (Dickie, Kerr, and Boudreau 1987). Thus, for example, the concept of ecological efficiency is defined by Slobodkin (1963) as the ratio of food intakes, $C_n/C_{n-1}$, by a predator-prey pair in an ecosystem. According to the foregoing schematic, ecological efficiency for the predator-prey pair considered in the corresponding biomass ratio can be written in parallel with the ratio of biomasses in equation 4.1 as

$$C_n/C_{n-1} = K_{n-1} F_n B_{n-1}/P_{n-1}, \qquad (4.2)$$

where the diagram, employing the principle of similarity of predator-prey interactions throughout ecosystems, shows that the efficiency terms involved in this ratio are those that apply over a different energy transformation phase of the overall system. The scheme thus allows us to see the impact of different values of parallel parameters taken over different phases of the system. If there were no differences in these efficiency terms between predators and their prey at different levels in a food chain, the ratios of biomasses and catches would have identical values. Conversely, observed differences between these ratios provide us with a way of evaluating the relative importance of the conversion factors within different parts of the ecosystem.

Similarly, according to the schematic, production efficiency, which is defined as the ratios of productions, $P_n/P_{n-1}$, between a predator and its prey, may be written for the same predator-prey pair as

$$P_n/P_{n-1} = K_n F_n B_{n-1}/P_{n-1}. \tag{4.3}$$

It is clear that differences between the production ratios and the biomass and catch ratios may be used in complementary fashion to study differences in the efficiency terms between successive steps in an idealized food chain system.

With this terminological clarification, it can be seen that the major concerns of energy transfer in food chains are all comprehended and can be explicitly measured in terms of the parameters of the biomass density spectrum, through the recognition of a given body-size ratio between a predator in a given trophic position and its prey in another position. Furthermore, all of these energy transfers are evidently related to the same three basic body-size-related parameters: the efficiency of energy transformation ($K$), the rate of predation ($F$), and the specific rate of production ($P/B$). While these three parameters are not necessarily independent of each other, they are clearly measurable as separate processes in the various predator-prey linkages that can be identified. Their common body-size dependence was the principal theme of Dickie, Kerr, and Boudreau (1987).

Clarifying concepts of energy transfer in the organism assemblages that underlie production in aquatic systems allows a renewed interpretation of both old and new observations of the numerical and biomass densities of organisms by body size, and the development of theory relating it to the underlying predator-prey energy transfer mechanisms. Such a theoretical foundation also makes it possible to extrapolate from scientific observations of the spectrum to the management of exploitation of natural aquatic biological systems, either through direct exploitation, as in fisheries, or through the perturbations caused by other forms of environmental disturbance. As shown in chapters 2 and 3, increased knowledge of various spectra or partial spectra has provided many new observations that have confirmed characteristic integral patterns affirming the existence of the spectrum as a general phenomenon. At the same time, others have exposed new structural features that appear to depart in details from these first generalized perceptions. For a comprehensive general biomass spectrum model, features at both the integral and the analytic scales need to be evaluated and, where appropriate, taken into account by a redevelopment of the theory.

## Formulating a Predator-Prey Model: Terminology and Definitions

Mathematical models of the spectrum of the biomass of organisms in relation to body size will vary in their assumptions and complexity of expression. However, all may be considered to begin from the observation of Sheldon, Prakash, and Sutcliffe (1972) that aquatic biomass density can be described by an allometric relation to body size. Following the terminology of Platt (1985), we may write it as

$$B_\nu \propto w_\nu^{-a}, \tag{4.4}$$

where biomass of a body-size group $\nu$, $B_\nu$, is expressed as a function of body weight $w$ for organisms within the size group $\nu$ having the nominal value $w_\nu$. In this formulation, $a$ is the allometric coefficient of the relation between body size and biomass revealed by the spectral plots.

As was indicated above, the term "biomass" has itself been used in various ways in the literature. Sometimes it is taken as an index of the *total* biomass of some class of organisms in an ecosystem. Alternatively, especially in plankton sampling, it is taken to be the biomass *density* of a given body-size class within an ecosystem. Throughout this book we shall define biomass as a measure of the density per unit area of a particular mass of organisms or its energy equivalent averaged within a defined part of an ecosystem.

In what follows, body size will always be reported as body weight in mass units or its equivalent in kilocalorie units. The area over which averages are taken is calculated from the projection of the water volume and the organism-containing bottom substrate onto the overlying, two-dimensional surface. That is, in general, we define the biomass spectrum for a unit area as the mass or energy equivalent of the biomass density of all organisms contained in each of a sequence of increments in the logarithm of the body size, within body-size limits set by the methods of sampling. In some cases investigators have calculated separate partial

spectra for pelagic or benthic parts of the system. Examples were given
in chapter 2: for instance, figure 2.1 is a biomass density spectrum for
the pelagic system. Spectra are frequently specified according to the water or bottom layer included in sampling.

In accordance with the descriptions of the biomass spectrum in
chapter 3, the designated body-size interval used to classify body-size
compartments, $dw_\nu$, will also generally appear to fall within one of the
series of trophic groupings or positions, n. The complete spectrum will
thus include the density distribution, by small body-size class intervals,
of all the organisms found in the defined sample area of sea or freshwater body.

Strictly speaking, a biomass spectrum is constructed from data taken
to be representative of some particular state of the system; formally the
spectrum is not a function of time. In practice, there is both a spatial
area and a time period of sampling over which an average biomass density can be estimated. This is, of course, determined by the usual considerations of representative sampling of dynamic systems. Often, the
time scale considered necessary to typify an aquatic system is simply
taken as the sampling interval itself. More often an average is calculated,
usually the annual average of the set of observations taken from time to
time over the principal production period in a seasonally varying sequence. The relevance of particular choices is evident from interpretations of spectra discussed in chapter 3. Workers interpreting the energy
flows according to both continuous and discontinuous models have
sometimes been concerned to look at the stability of the biomass domes
through time. In some cases, particularly with continuous models, it has
been deemed important to detect the extent of possible progressive
changes in the domes, such as might be anticipated from growth and
mortality of the constituent organisms. In the succeeding discussions of
data supporting theoretical predator-prey models, we will initially consider the spectrum to be stationary in time. Developing time-dependent
models is a specific additional undertaking, which is discussed at appropriate points.

Mathematically it is often convenient to follow the continuous spectrum approach by considering that $B_\nu\, dw_\nu$ is the increment of biomass
density corresponding to the body-size increment $dw_\nu$ in trophic position n. This definition of $B_\nu$ or spectral density function, in which biomass density is expressed "per unit body size," was called the *normalized*

biomass spectrum by Platt and Denman (1977, 1978). Its convenience rests in the fact that because $B_\nu dw_\nu$ is always divided by the corresponding $dw_\nu$, the resulting spectrum is always independent of the sometimes varying widths of the body-size intervals employed. The continuous normalized spectrum can conveniently be used to describe and compare communities of aquatic organisms over a wide range of body sizes, ideally where the numbers of organisms and organism sizes are sufficiently large that $w_\nu$ can be considered a continuous variable rather than a set of discrete integers.

Strictly speaking again, however, $w_\nu$ cannot be continuous for any finite ecosystem. For this reason, as well as for purposes of illustration, it is also often convenient to express the logarithm of the biomass density in equal intervals of log body size in what we have already called the *unnormalized* spectrum, much like that first used by Sheldon, Prakash, and Sutcliffe (1972). To distinguish the two we also define the unnormalized biomass spectral density function, $\beta_\nu$, where it is understood that this form of the spectrum must be presented in equal logarithmic intervals of body size, $\Delta w_\nu$, centered on $w_\nu$. That is, the parameters of the unnormalized spectrum are not independent of the body-size classification employed, and so it needs to be standardized for comparative study. It should be clear that for any $w$

$$B_\nu(w) = d\beta_\nu(w)/d(w). \qquad (4.5)$$

Both normalized and unnormalized forms of the equations have their preferred uses. Thus, if we employ the normalized spectral definition given by equation 4.4, a plot of the logarithm of normalized biomass, $B_\nu$, as ordinate, against log $w_\nu$, as abscissa, may be considered to yield a straight line with slope $-a$ (figure 2.3). That is, the normalized plot offers an immediate estimate of the value of the allometric coefficient. Furthermore, as pointed out by Platt and Denman (1978), this slope, in the $\log_2$ or octave units most often used to display requisite detail in the spectrum, has the added convenience of being the same no matter whether the abscissa is plotted in biomass or numerical density terms per unit body-size interval.

The spectral data summed over discrete, rather than continuous, logarithmic body-size intervals, $\Delta w_\nu$, following the unnormalized spectrum, $\beta_\nu$, also fall along a more or less straight line such as was drawn by Boudreau and Dickie (1992) and given in figure 2.4. In this case, however, according to equation 4.5, the unnormalized spectrum will have a slope $-a + 1$.

Because the value of $a$ is often found to be very close to $+1$ for aquatic ecosystems, the visual profile of the discrete spectrum tends to be close to the horizontal. This fact simplifies the problem of scaling of comparative illustrations over a wide range of body sizes and is often employed because of this. In chapter 3 we discussed instances where partial spectra show a positive unnormalized slope, and others where the slope is negative but close to the horizontal. These low slopes are often called the slope of the spectrum, although for obvious reasons they should be understood as the slope of the unnormalized spectrum, to ensure their distinction from the allometric coefficient.

We can also construct spectra of production in the ecosystem. Hence, in parallel with the biomass spectral density functions, we may define the production spectral density functions: $P_\nu$ in parallel with $B_\nu$, and $\pi_\nu$ in parallel with $\beta_\nu$. The production spectrum is also allometrically related to body size (Dickie, Kerr, and Boudreau 1987). Whence

$$P_\nu \propto w_\nu^{-a''}. \tag{4.6}$$

In the descriptions of the spectrum in chapters 2 and 3, we have also been interested in the specific production, which, as the ratio of two allometric functions, itself may be expressed allometrically as

$$P_\nu / B_\nu \propto w_\nu^{-a'} \tag{4.7}$$

where it is evident that

$$a = a'' - a'.$$

With this agreed terminology, we are in position to consider various possible models of the underlying dynamics.

## Two Predator-Prey Equations

Aquatic ecosystems may be generally considered as energy flow systems in which the energy derived from the sun flows through a spectrum of biomass classes, from primary producers through small body-sized predators that act as prey for successive groups of larger predators. That is, the flow takes place through a series of distinct body-size steps or jumps that make up the trophic linkages in a chain of successive prey and predator relationships. In envisioning such a system, we are also invoking the principle of biological similarity between the various trophic linkages. In such a system, predators are unlikely to act arbitrarily. Rather, as we have indicated at several points, we take the position that predators will act somewhat in common; most of the predators in any one class consume a given, though not necessarily small, size range of prey, but act on these prey in much the same way as their own predators act on them. In most cases we shall assume that the body size of prey will be smaller than that of the predators. Furthermore, we expect that the ratio of body sizes between predators and prey will have a basic similarity throughout most of the system. Our interest in a mathematical spectral representation of such an idealized ecosystem revolves around its potential for more precisely reflecting the characteristics of transmission of the total food energy when it is dependent on linkages that have developed over long periods among the successive predator-prey pairs that make up natural ecosystems. We therefore begin our modeling of the energy transmission using these simplifying though not unreasonable assumptions. The effects of relaxing them will be considered later.

In this section we consider two predator-prey models through which this energy flow has been represented by model builders. In the following section we will consider how some of the empirical results outlined in chapter 3 imply spectral features that distinguish both of these specifically predator-prey models from the general mathematical, continuous energy-flow model proposed by Platt and Denman (1977, 1978).

We begin by developing a representation of energy flow in the form of linkages between any two members of a predator-prey pair. We call this Model I. In keeping with the empirical plots of chapters 2 and 3, we initially envision these feeding links as occurring between successive trophic "positions" in the spectrum. That is, we shall at first consider a model in which organisms in one of the size compartments in a dome-shaped biomass grouping or position labeled "n" feed on organisms within one size compartment in the next smaller dome-shaped biomass grouping, which we have designated n − 1 as in figure 4.1. Thus, we consider each dome-shaped grouping to be a functional trophic grouping, and have labeled it a "trophic position" as defined in chapter 3 (Levine 1980).

In chapter 5, using Model II, we will consider the question of feeding linkages in more detail, allowing links within a trophic position as well as between them. Relaxing the unrealistic assumption that one size class of predator can feed on only one size class of prey in another trophic position has some important consequences for developing the equation systems. The initial restrictions of the definition of predator and prey made here are only to clarify the potentially greater complexity of mathematical symbolism required to describe an entire structural assemblage.

We label the initial prey and predator groups as $i$ and $j$, respectively. Members of any particular trophic position, $j$, initially considered as predators, may have similar feeding habits and diets, but may, in fact, be composed of several size classes, each having different predatory habits, and may feed within a limited but not necessarily small range of body sizes of their prey. They may, in their turn, be prey for a group of larger predators. We initially consider the food linkages in the model as repeated simple predator-prey chains. These simplifying assumptions are not critical. We develop our arguments using them so that later we can be explicit about the effects of the various specific ways in which they can be relaxed.

First, we direct attention to the allometric equations, 4.4, 4.5, or 4.6, above. If we consider an energy flow system like that of figure 4.1, characterized by predator-prey linkages displaying simple expressions of similarity, we can write the ratio of productions between a given set of predators and their prey as

$$P_j/P_i = (w_j/w_i)^{-a''}.$$

Taking into account the allometry of equation 4.7, and remembering that different values of $\nu$ refer to values of $i$ for the prey and $j$ for the predator, we can write a biomass density function for Model I as

$$A_j/A_i = (w_j/w_i)^{-a}, \qquad a = a'' - a' \qquad (4.8)$$

where we have written $A_\nu$ for the biomass spectrum, in place of $B_\nu$, as a reminder of the purely *A*llometric basis for this first simple model.

This is the form of biomass spectrum model proposed by Borgmann (1987), who was interested primarily in various ways of approaching sampling and interpretation of the allometric exponent, $-a$. Implicit in this formulation, as pointed out by Borgmann, is that all food energy passing through the group labeled $i$ later appears in the group labeled $j$. This approximation to the actual dynamics of an ecosystem may be useful for a variety of purposes, although for other purposes (e.g., in defining the slope of the spectrum as a simple, straight line) it imposes an unrealistic limiting condition on the model.

A different approach was proposed by Boudreau, Dickie, and Kerr (1991), based on perceptions of body-size dependency developed by Dickie, Kerr, and Boudreau (1987) and incorporated in figure 4.1 describing the community algebra. This model, which we examine in detail in this book, is based on the explicit assumption that the production of the predators in feeding size group $w_j$ is dependent on the biomass density, or the equivalent specific production of the prey in size group $w_i$, but that there will be energy losses from the system at each step in transmission and conversion. Thus, we also invoke an energy conversion efficiency that in equation 4.1 we conceived in two parts. The first part of this conversion efficiency term, now labeled $K_j$, describes the efficiency with which these predators convert an acquired unit of prey biomass to predator biomass. The second part, which we label in this more specific development as $F_j$, is taken to represent the fraction of the biomass density of the prey in the size group $w_i$ that is captured by the predators in size group $w_j$. That is, this model is different from that described by Borgmann on two accounts. First, it takes account of losses of the total energy in the system, both during its transmission and during its metabolic conversions. Secondly, it specifically makes the production of the predator dependent on the density of its food resources, the prey density.

We believe that these qualifications about the efficiency of energy transfer need to be made explicit in a model for several reasons. In particular, we wish to point out that while empirical representations of the spectrum in chapter 2 show that the overall or primary slope of the biomass may often be indistinguishable from a straight line, the discussions and descriptions in chapter 3 indicated that this is certainly not always the case. There may be a number of underlying causes for this departure from a simple straight line (which we will discuss later). For example, even in connection with the energy conversion terms, there is good reason to suppose that the fraction of a given prey size captured by predators, $F_j$, may be quite different for different sizes of predators; zooplankton predators appear to take a higher fraction of the average biomass of their prey than do fish or whales (Dickie 1976). If $F_j$ varies, representing the spectrum by a straight line would then imply that there must be a single, major, predetermined set of compensatory changes in the efficiency of conversion, $K_j$, over the same body-size ranges of all organisms. These changes have been extensively studied in fishes (Paloheimo and Dickie 1966b), where they have been found to depend on body size of prey as well as the predator-prey body-size ratio. Such effects may thus effectively give rise to curvatures in the spectrum, without even invoking complexities that must arise from differences in the density distributions or behaviors of prey or predators. We discuss the implications of these and other such factors in more detail in chapters 5 and 6. For the present we need only note that identification of components in a conversion term allows us to explicitly examine the effects of evident differences in feeding strategies among predators in relation to more general expressions of the energy flow in the spectrum.

Phenomena relating both the relative densities and the relative body sizes of predators to their prey will be important to the rates of flow in ecosystems on various time and space scales. On the longer, evolutionary time scales, for example, the trophic flow through any ecosystem must depend on the development of the particular physiological and behavioral mechanisms of the combinations of species that function as its carriers (cf. Warwick 1984). The apparent stability of the overall result in communities consisting of quite different species has occasioned wonder at the stability and ubiquity of the apparent system properties of the spectrum and led to the theories of energy transmission that we are discussing.

It is evident, however, that compensations of both energy uptake and the efficiency of its use for production must also take place on the shorter time scales, such as on the ecological scales of adjustment in organism density. The time scales of density changes must be sufficiently short term to ensure that predators can satisfy their day-to-day, individual food requirements; that is, within any naturally occurring feeding assemblage, the predators must adjust their local density so that the energy intake from the rations obtained by an individual predator will be in balance with its "routine" metabolic requirements (Dickie, Kerr, and

Boudreau 1987). From this requirement it follows that the specific production of the predator must be an important short-term function of the density of its prey.

These facts about the feeding linkages can be recognized by utilizing the conversion coefficient in the form

$$P_j/B_j = K_j\, F_j\, B_i/B_j. \tag{4.9}$$

If we recognize the allometric nature of the specific production, on the left-hand side, as in equation 4.7, and assume that the conversion term of the right-hand side can also be described in allometric form, this equation allows us to write, in parallel with equation 4.8, the biomass density ratio for Model II:

$$B_j/B_i = \alpha w^\gamma, \tag{4.10}$$

where by rewriting the allometric relation of biomass to body mass in the form familiar from metabolic work, $w^\gamma$, we emphasize its necessary relation to both the allometric coefficient and, from the discussions in chapter 3, its twofold scaling.

We justify writing the conversion term in allometric form in chapter 6. Here, we wish only to emphasize that while the two models described in equations 4.8 and 4.10 have different right-hand sides, which invite the study of different biological mechanisms, the left-hand sides are identical. We have used different terminology in the two models to emphasize that there are the two distinct sources of assumptions about the precise forms of the energy transmission from which the observed biomass spectrum may be considered to arise. Initially, however, the generality of the theory underlying this predator-prey modeling exercise most depends on the solutions of the identical left-hand sides of both models. We therefore direct attention particularly to solutions for the left-hand sides of the two equations.

## Solutions of the Mathematical Predator-Prey Equations

The formulations of the two equations of energy flow in ecosystems according to equations 4.8 and 4.10 are deceptively simple; the conditions they imply, however, have a far from trivial effect on the apparent energy flow. The predicted results can thus be fruitfully compared with the empirical findings. It is the common structure of the solutions in this form that also gives the models we describe here characteristics that distinguish them from the continuous energy-flow model proposed by Platt and Denman (1977, 1978). We need to consider the commonality and the differences between these fundamentally different approaches.

In common with our approach, Platt and Denman recognized the potential power of the allometric representation as a means of generalizing the mathematical structure of the biomass spectrum and its relation to ecological mechanisms. However, they based their development on the visual representation of the large-scale plot of log biomass density on log body size, from bacteria to fish. They further accepted that this relation could be expressed as a straight line of negative slope, and so embodied in their model the concept of a steady flow of energy through contiguous population elements along an essentially smooth, linear array of population biomass elements. That is, while they derived a model solution that was stationary in time, as is ours, this was regarded as an approximation to a model in time, in which there might be local variations of higher and lower biomass densities, but not in any fixed position in relation to body size. From this point of view, the sloping spectral line was rather a locus of biomass densities along which the biomass population elements would be expected to exhibit a relatively smooth net flow from small to large sizes. Although there could be complex internal movements of energy both up and down the body-size classes at finer scales of observation, their averages were expected to result in a smooth spectrum. Empirical tests of the validity of this model were discussed in chapter 3.

Our reformulation, of discrete jumps in energy between the body-size elements constituting relatively fixed body-size ranges of prey and their predators, leads to a distinctively different mathematical structure of the energy flow in the spectrum. As will be seen below, the model proposed by Borgmann (1987) has much in common with that proposed by Dickie, Kerr, and Boudreau (1987), and by Boudreau, Dickie, and Kerr (1991). All have the same predator-prey jumps in energy flow, and, as a result, show distinctive properties that differ from those of continuous flow models.

The customary interpretation of the normalized biomass spectral density function, common to all the models, is that biomass density is a function of body size. In this interpretation, it is also conventional to imply that the function is single valued. That is, for any argument (= body size) in the domain of the function, there corresponds one and only one value of the function (= biomass spectral density). While distinctions between single- and multiple-valued functions may at first also appear to be trivial, particularly in a continuous flow model, we again emphasize that, while the single-valued function makes initial solution of the equations more evident, it also places unnecessary restrictions on the biological interpretation of the spectrum as a predator-prey model. In this sense, too, the predator-prey models allow a representation of the model system that appears to be both more flexible and more realistic.

For example, a single-valued representation of the organization of an ecosystem composed of predators and prey implies that all predators are larger than all prey. This is unnecessarily restrictive, even for an aquatic system in which that seems often to be the case. In what follows, therefore, we initially consider the single-valued function and its implied *single* spectrum. In the next section we shall consider the practical implications of an alternative multispectrum interpretation in which this relative body-size restriction can be relaxed.

For the first, single-valued analysis we write

$$A_\nu = A(w_\nu), \qquad B_\nu = B(w_\nu), \qquad \nu = i, j. \qquad (4.11)$$

From the definition of these spectral expressions in equation 4.8 and 4.10, it can be seen that the numerators of the left-hand sides of the ratios are simply the spectral density function evaluated over the range of predator body sizes. The denominators are the *same* spectral density functions evaluated at the (nonoverlapping) prey body sizes. As a means of simplifying the equations, as well as of identifying properties relating these two groups of organisms in the spectrum, we may define the required predator-prey body-size ratio as

$$R = w_i/w_j, \qquad (4.12)$$

which we will consider to be characteristic of any predator-prey linkage and to be a property of the predator. With this application of the principle of similarity we can thus simplify terminology, dropping subscript *j*, and letting $w_i = Rw_j$. We can also define a *size-interval* ratio, $\Delta w_i / \Delta w_j$, as a size concentration factor that would permit us to study situations in which grazing by the size range of the predator may show different relative ranges of prey size. In the mathematical development of the model, however, we need concern ourselves only with the size ratio, *R*. Initially we shall consider *R* to be constant and to have an upper bound considerably less than 1. We initially also assume that the size-interval ratio takes a value of 1.0. Later we will discuss the effects of relaxing these stipulations.

Substituting equation 4.11 and 4.12 in equation 4.8 and 4.10 allows us to write for the two models

Model I
$$\frac{A(w)}{A(Rw)} = R^{-a} \tag{4.13}$$

Model II
$$\frac{B(w)}{B(Rw)} = \alpha w^{\gamma}.$$

It has been pointed out by Thiebaux and Dickie (1992, 1993) that a solution to these equations becomes more apparent if the equations are recast in terms of the logarithmic variables used to plot the data in biomass spectra. That is, in Model I we let

$$\log A(w) = y(x), \text{ and } \log w = x.$$

Model I can then be rewritten as a log-log plot in the form

$$y(x) - y(x + \log R) = -a \log R. \tag{4.14A}$$

With parallel substitutions for $B(w)$ and $w$, Model II can be rewritten as

$$y(x) - y(x + \log R) = \log \alpha + \gamma x. \tag{4.14B}$$

Once again it is apparent that the left-hand sides are identical formulations, which can be seen mathematically as difference equations whose solution $y(x)$ is the sum of a particular function related to the specific right-hand sides, plus a periodic function related to the value of *R*.

Returning to the biomass spectrum terminology of the two models, we may then write for

Model I $\qquad \log A(w) = -a \log w + H_0 + H(\log w)$ $\qquad$ (4.15)

Model II $\qquad \log B(w) = \frac{1}{2} c_0 [\log w/w_0]^2 + J_0 + J(\log w).$

This comparison shows how the first term of the solutions to the two equations depends on the form of the biological functions incorporated into the modeled energy transfers. In Model I the first term is a straight line of slope $-a$, and a free intercept that may be taken as having the value of the second term, $H_0$. In Model II, the first term is a quadratic with curvature $c_0$ and vertex at the point $(\log J_0, \log w_0)$. That is, ignoring the periodic part for the moment, the first parts of the solutions to the two model equations suggest that the biomass of assemblages of predator-prey pairs either will be in the form of a straight line or will form a background parabolic dome. It may, of course, have a curvature low enough over the range of observed body sizes to be indistinguishable from a straight line.

The effects of different values of $R$ are directly evident in the equation expression of the log-log plot for Model I in equation 4.14A. The detailed equations for Model II, which can be derived from equation 4.14B, show that

$$c_0 = -Y/\log R, \qquad w_0 = \sqrt{R}/\alpha^{1/\gamma}. \qquad (4.16)$$

That is, both of these parameters are also related to the predator-prey body-size ratio, $R$, as is directly evident in the equation for Model I (equation 4.14A). We note in passing that the curvature, $c_0$, which is the second derivative, $d^2y/dx^2$, must be negative in order for the biomass to be finite within the bounded biomass spectral density. As we shall see below, this seems to be the case, since $\log R$ is negative, as are most values of the population $\gamma$. However, even if this were not the case, we would need to recognize that we are not attempting to deal with an infinite system. The whole spectrum is truncated at the sizes of the smallest measurable microorganisms and the largest whales. Therefore, slight positive curvatures would not be strong reasons for rejecting the mathematical formulation.

The common and unique nature of the mathematical solutions to these predator-prey formulations is, however, shown by the second and third terms of equation 4.15, where the expressions

$$H_0 + H(\log w) \quad \text{and} \quad J_0 + J(\log w)$$

are, in the sense that they are specified only by factors outside the mathematics of the equations themselves, arbitrary periodic functions of $\log w$, having a period $|\log R|$. That is, the mathematics shows that superposed on the smooth background line or parabola, there will be a repeated series of oscillations in biomass in relation to body size, in which the residual variation will show some multiple of the period $|\log R|$.

According to this formulation, oscillations of biomass within given trophic groupings would be stable within a simple model in which the average feeding size-ratio $R$ had no variance. Later we discuss results of variations in $R$. At this point, however, it should be noted that the fact that the model predicts oscillations in biomass, but can really say nothing more about them, should not be taken to mean that they are unimportant effects of negligible influence. The observed effects will depend on the scales of observation. In general we can say that they will appear as stationary perturbations imposed on the overall spectral curve of biomass density with body size, hence will have shorter periods than those characterizing the dominant biomass parabola and reflected in the quadratic term.

This periodicity arises solely from the nonlocal nature of the energy exchange in the predator-prey system, and may be affected by a number of factors intrinsic to the predation system. It may even be related to extrinsic environmental factors if these influence the body-size distributions of predators and their prey. But in any case, the nature of the oscillations is such that if they did not appear in data generated by predator-prey interactions, the lack of oscillation around the basic, smooth linear or quadratic integral base would need to be explained.

In a general way, oscillation in the actual profile of the biomass spectrum has always been seen in biological oceanographic sampling, where it takes the form of the characteristic stationary body-size groupings of phytoplankton, zooplankton, benthos, and fish. These groupings are so stable that they have had different special sampling gears developed to study them. As we have seen in chapters 2 and 3, available data collections suggest that stationary domes are, in fact, a prevalent feature of aquatic ecosystems. It is on this basis that we pursue the consequences of relaxing some of the assumptions implicit in this first, somewhat simplified, representation of the predator-prey energy flow system.

## The Multispectrum Alternative Interpretation

The mathematical difference between a single-valued and multiple-valued system is profound. In addition, there are a number of biological situations in which the assumptions of a single-valued spectral function are not satisfactory. The possibility of some predators being smaller than some prey has already been mentioned. But reservations about the precise assumptions underlying the model may extend to more fundamental issues than that. For example, we may question the immediate trophic interdependence of the successive stationary domes that appear in the empirical spectrum for some very broadly sampled areas. Similarly, we may have questions about the strength of the competitive interlocking links between apparently parallel food chains in a system composed of several, more-or-less independent predator-prey food transfers. That is, the distinction between the single-valued and multiple-valued spectrum may not be trivial from either a biological or a mathematical point of view.

Our interest here is to show that we can usefully apply the basic methodology of the spectrum to describe the properties of a collection of spectra, without committing ourselves to the postulation on mathematical premises of a degree of system interaction that cannot be established from the field evidence.

To make this multispectral definition clear, we may write in place of equation 4.11 the notation

$$A_\nu = A_\nu(w_\nu), \qquad B_\nu = B_\nu(w_\nu), \qquad \nu = i, j. \qquad (4.17)$$

If we use this interpretation to examine equations 4.8 and 4.10, it should be clear that in this case the numerator of the left-hand side of each of them is a predator spectrum evaluated at the predator body sizes. But now the denominator is a different spectral function, evaluated at the prey body sizes. The spectral representation of the sequences of body sizes, in which there are either predators or prey, is now merely used to

measure the extent to which two or more spectra have similar characteristics and to calculate parameters of slope and period for the combined spectra.

The resulting models take the same basic structural form that we
have seen above. That is, we may still define a predator-prey body-size
ratio $R$ and define $w_i = Rw_j$. Substituting equation 4.17 in equations
4.8 or 4.10 then gives

Model I
$$\frac{A_j(w)}{A_i(Rw)} = R^{-a} \tag{4.18}$$

Model II
$$\frac{B_j(w)}{B_i(Rw)} = \alpha w^\gamma$$

If we again translate these equations into the variables used to plot the
log-log data, and define $x = \log w$ and $y_\nu = \log A_\nu$ or $\log B_\nu$, we arrive
once more at difference equation forms:

Model I $\qquad y_j(x) - y_i(x + \log R) = -a \log R \qquad (4.19)$

Model II $\qquad y_j(x) - y_i(x + \log R) = \log \alpha + \gamma x,$

which have solutions parallel with the solutions found for the single-
spectrum interpretations. That is, the solutions each consist of two
parts: a specific part related to the right-hand sides of the two equations,
plus a periodic part having a period that is a multiple of $|\log R|$, but
otherwise not specified within the mathematics.

In this multispectrum interpretation we have indicated by the equation for Model I that the predator spectrum is identical in shape to the
prey spectrum, but shifted to the right by the amount $|\log R|$ and down
by the amount $a|\log R|$. This severely limits the interpretations that can
be allowed in this simple, possibly oversimplified form.

In Model II the predator spectrum must in general be related to the
shape of the prey spectrum shifted horizontally and vertically. But in addition to the shifting, the presence of the linear term $\log \alpha + \gamma x$ gives
rise to a distortion in the shape of the predator dome, which may be
thought of as a vertical linear shear. Mathematically, however, since the
domes are both parabolas with their axes parallel to the ordinate, that is,

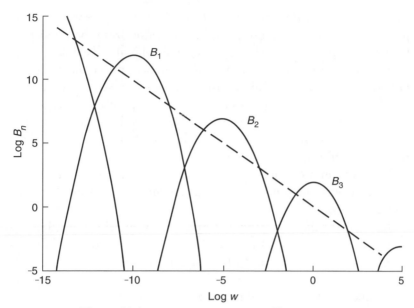

FIGURE 4.2 The multiple spectra interpretation of biomass size spectra. As displayed here in log-log coordinates, the same shape is equally spaced horizontally and vertically. (Redrawn from Thiebaux and Dickie 1992.)

parallel to the shear, they are merely displaced vertically without shape distortion. As a result, the parabolic term of Model II introduces an effect on the estimate of the slope of the spectral line derived from the succession of predator-prey spectra, but does not otherwise affect the fittings.

From the mathematical point of view, the multispectral interpretation is quite different from the single-valued interpretation. In particular, it allows for the fact that there may be overlapping to an unspecified extent between the body-size distributions of the predators in one trophic position and their prey in another. We illustrate it in figure 4.2, where it can be seen that while any intermediate dome constitutes predators of the next smaller dome and prey of the next larger, there may well be some prey that are larger than some predators. Such considerations make it possible to obtain realistic representations of aquatic ecosystems where the body sizes of a class of small-bodied predators with high specific production may overlap the body-size range of the class of the largest and least productive prey.

However, from the practical point of view, it may be virtually impossible to distinguish the single-valued situation from the multispectral situation purely on the basis of an empirical plot of field data. That is, in complex communities or in situations of automated measurement of body size of the large numbers of organisms involved, it may not be possible to specify a priori to which trophic grouping a given organism belongs. This suggests that errors might be made in spectral interpretations of the organisms at the points of overlap in the size distributions in a multispectral situation. In such a case, whether or not the body sizes form a multiple spectrum, the observed spectral density at any body size can be considered only as the sum over all trophic groupings at that size.

Recognizing that the individual parabolas are actually normal distributions when plotted on the biomass scale instead of the log biomass scale, Thiebaux and Dickie (1992) examined this situation mathematically and demonstrated that in the infinite spectrum, the multispectral case reduces to the single-valued case, so that there is no error in estimating the important parameters of slope and spacing between domes. Since, however, the biological spectrum is truncated at the smallest and largest body sizes, some distortion is inevitable. The calculations of Thiebaux and Dickie (1992) indicate that in the usual case of the whole biomass density spectrum, where minima between peaks in the biomass density distribution are several orders of magnitude below the peaks, they can be effectively treated as zero-density gaps, and errors in parameter estimation will be negligible.

## Summary

We have developed a simple mathematical theory of the dynamics of energy flow in a model ecosystem organized as illustrated in figure 4.1. The flow is based on a pattern of similar predator-prey transmissions that convey the energy throughout the total community biomass density. The spectrum that emerges from this conception has a number of interesting properties that correspond with patterns observed in empirical observations of aquatic biomasses. In particular, the mathematical formulation leads to the expectation of two different time scalings of its functioning: at the broadest community scale there appears a characteristic overall or integral biomass distribution. Within this integral level, however, there appears a secondary structuring consisting of a series of stationary biomass domes periodically spaced along the integral profile. The spacing of the domes is a function of the relative body sizes of predators and their prey.

The formulation also leads to the expectation that while many observed integral spectra seem to have been fitted satisfactorily by a low, negative, straight-line slope, this is by no means a certain or obligatory feature. In fact, from the explicit algebraic formulation, it would be surprising if the strong dependence of predator production on both rates of predation and metabolism in relation to body size and densities of prey did not lead to curvatures in the general profile of the spectrum, of the sort predicted by Model II. As is evident from the discussions in chapter 3, the empirical evidence here needs a closer examination in relation to theory. Evidence for departures from a straight line are most obvious for the domes that describe particular trophic positions. However, curvature also seems to be a strong possibility for the integral spectrum as a whole.

Clearly, the relation between theory and the empirical results needs to be examined in detail if we are to be confident that this formulation of the dynamics captures the principal features that control the distribution of biomass and production within the spectrum of body sizes. The capacity of this simple predator-prey model to predict so many features of the empirical spectra induces us to continue our study of it in the next chapter with a more detailed discussion of the nature of the predator functions.

# CHAPTER 5

ॐ

# Feeding Relationships in the Spectrum

The models outlined in chapter 4 meet certain expectations of the observed general shape and structure of both the biomass density and production density spectra. However, a number of questions about their reality and completeness need to be addressed. No model can be expected to explain the history of an aquatic ecosystem or predict its future states unless there is reassurance that it reflects the essence of both the principal mechanisms controlling energy flow and the actual energetic interactions that underlie the structure.

The abstract simplifications of the original models do not meet the criteria of either reality or completeness. In the first place, and perhaps most importantly from the point of view of the realism of predator-prey models, the initial development described in chapter 4 incorporated simplifying assumptions about the nature of feeding relationships in aquatic ecosystems. The regularity in structure of the biomass density, echoing the underlying predator-prey interaction mechanism of the model, depends on our invoking the principle of similarity, from which we inferred a constant ratio of body sizes between predators and their prey throughout the food chains. While there is no reason to deny this principle, up to this point the models have dealt only with the simple transmission of food energy between trophic positions.

In the real world, there are differences in the ratio of body sizes between predators and their prey and there are feeding linkages within trophic positions. For example, large zooplankton feed on small zooplankton, and large fish feed on small ones. Thus, while the initial models take into account the fact that all food energy reaching higher trophic positions must jump the body-size gaps between trophic positions, feeding within positions means that energy transmission also involves body-size jumps within them. Jumps *within* trophic positions are not necessarily of the same magnitude in relative body size as those *between* positions. Therefore, the realism and completeness of the models can be enhanced by incorporating these additional common features of feeding interactions. We wish to ascertain the effects of relaxing the initial simplifications and to find the nature of modifications that will more nearly image these additional features of observed systems.

In the second place, and at the most basic level of the predator-prey equation system, it also needs to be established that the efficiency terms included in Model II can be incorporated into the spectral representations in such simple, allometric fashion. Establishing the allometric basis for the conversion coefficients is not necessarily a straightforward matter, especially when it is remembered that in the equation systems under discussion we must be able to deal with at least the twofold physiological and ecological scaling of the underlying dynamics. The twofold scaling that was initially most evident in plots of specific production must also affect the efficiency terms that are a part of the detailed biomass spectrum equations.

In this chapter, we examine the effects on the model of elaborations in the feeding strategies in relation to the posited twofold scaling of the dynamics of production. As we shall see, refinements and elaborations of the mathematical statements to incorporate the kinds of ecosystem interactions that have been observed result in predicted configurations that are compatible with the special conditions adopted for the initial explication. The development thus verifies that the predator-prey model elaborated as Model II meets the most basic of the requirements of reality and completeness. Questions about the role and the values of the production coefficients are reserved for examination in chapter 6.

## Fitting the Density Spectra at Different Scales

To this point, we have studied density spectra in two forms: the biomass density spectrum and a corresponding production density spectrum. Empirical studies of both spectral forms have revealed an overall or integral pattern as well as a persistent internal structure, depending on the degree of aggregation and classification of the observed data. In the course of discussing detailed fittings of more elaborate models, it is useful to review the characteristics that have already become apparent at the two scales.

## *The Integral or Primary Scaling*

The twofold scaling of the underlying dynamics was initially identified in relation to the specific production spectrum. Dickie, Kerr, and Boudreau (1987), using the analyses of Humphreys (1979) and of Banse and Mosher (1980), found that the allometric slope parameter estimated from production density data over all trophic positions took a value of about $-0.18$. This corresponds to an allometric slope of total metabolism of 0.82, which is within the range of values found in physiological studies of whole organism metabolism (Fry 1957; Hemmingsen 1960). Because this general slope of the production spectrum appears to primarily reflect the changes in metabolic rates of individual organisms with body size, Dickie, Kerr, and Boudreau (1987) called this method of calculating parameters of the spectrum a "physiological" scaling.

Banse and Mosher (1980) recognized that a higher-valued negative slope is found when the specific production data are compiled within trophic positions. This slope takes a value between about $-0.33$ and $-0.37$, a higher value than can be attributed to the metabolism of individual organisms. In accordance with interpretation of the compilations by Humphreys (1979), the slope appeared to depend on the aggregation of data to represent average metabolism of the densities of organisms of different sizes per unit area of habitat. That is, the higher slope within the trophic positions reflects not only the changes with body size in the metabolism of individual organisms, but in addition, adjustments in the area-aggregated metabolism represented by the changes in predator density necessary to permit development of trophically related communities. Because the higher slope reflects these additional ecological adaptations of aggregations of organisms, Banse and Mosher called it an "ecological" scaling.

Figure 5.1, redrawn from Boudreau, Dickie, and Kerr (1991), is designed to clarify the concepts underlying these two scalings in relation to models of both the production density and biomass density spectra. The upper panel, figure 5.1A, shows a schematic plot of the unnormal-

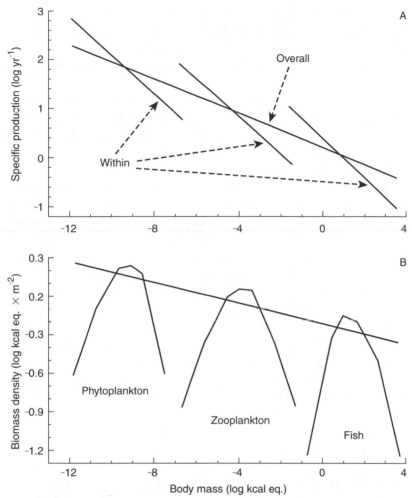

FIGURE 5.1 Schematic representation of the equivalent body-size spectra of (*A*) specific production, expressed as log *P/B* by trophic positional groupings, and (*B*) biomass density of a given environment. (Redrawn from Boudreau et al. 1991.)

ized production density spectrum. There is an overall line of low nega-tive slope representing the physiological scaling of production density. It is intersected by parallel lines of higher-valued negative slope, the eco-logical scalings, which span the body-size range of the organisms mak-ing up each of the trophic positions. These shorter and steeper lines might well overlap in body size at their ends if the data collection had been able to distinguish predators from prey at all body sizes. That is, the data fittings could employ the multiple-valued rather than the single-valued solutions to model equations described and discussed in chapter 4.

The lower panel, figure 5.1B, shows the corresponding schematic, unnormalized, integral biomass density spectrum. In this case, the physiological scaling that applies to the whole spectrum again shows up as a nearly straight line of low negative slope. Clearly, its slope value (or low curvature) is directly related to the overall slope of the production spectrum. However, the relative statistical reliability of fittings of production and biomass spectral data poses problems, as do the values of the efficiency coefficients, so that exact numerical equivalence is not easy to establish a priori. As was suggested by the empirical plots in chapter 3, the unnormalized slope of the integral biomass density spectrum often appears to be very close to zero; that is, it may have characteristics very close to the "flat" spectrum originally described by Sheldon, Prakash, and Sutcliffe (1972). Estimating parameters for the equations will be studied later.

The integral biomass density slope may be drawn through the averages of body size and biomass density calculated for each of the various domes that make up the secondary structure of the spectrum. Each dome, representing a trophic position, generally corresponds with one of the familiar designations of secondary structure: the production density groups referred to as bacteria, phytoplankton, benthos, zooplankton, and so on. As with the production spectrum, overlaps in the body-size range between biomass density domes depend on the collector's ability to distinguish the trophic position of the organisms found at all of the body sizes "intermediate" between domes. The approach to a straight line, or the degree of curvature, depends, of course, on the solutions to the equations fitted to the data.

In fitting the data on biomass density to the solutions to the model equations (4.15) in normalized form, Thiebaux and Dickie (1992) initially treated the physiological scaling as the primary object of analysis. That is, according to either Model I or Model II versions of the equations, the overall structure of the spectral biomass density would be either a straight line of slope $-a$ ($[-a + 1]$ in unnormalized form), or a shallow parabola with its apex position at ($\log J_0$, $\log w_0$), where $J_0$

is the intercept of the overall line and $w_0 = \sqrt{R}/\alpha^{1/\gamma}$. Curvature at the apex of the dome is denoted by $c_0 = -\gamma/\log R$. On the basis of many empirical fittings of biomass spectra, we would expect the curvature to be relatively small for the integral spectrum, in some cases so nearly a straight line that it might be difficult to distinguish the relative goodness of fit of Model I or II from the empirical data alone.

However, in both Models I and II the initial solutions lead to the expectation of oscillations in the biomass density with body size along the integral slope. It is these oscillations applied to the entire data set that appear as the domes or parabolas of what we have termed the trophic positions. That is, the oscillations in biomass at successive body-size classes define a secondary scaling that results in the secondary structure of trophically related ecosystems. Evidence for the secondary structuring would be expected to appear in the fitting of data to the integral spectrum.

According to the equation system, the spacing between the vertices of these biomass domes of the secondary structure is a function of the predator-prey body-size ratio between predators in one dome and prey in the next smaller one. We have written it as $|\log R|$ in the equation system. Inspection of equation 4.14 and equation 4.16 shows the number of ways in which $|\log R|$ influences the values of the positions and curvatures of the repeated domelike structures within the integral slope in the equations of Model II. This clearly indicates the importance of assumptions about the average ratio of body size between predators and prey in successive trophic domes for determining the precise form of this secondary or ecological structuring in the model.

### The Secondary Scaling

It should be evident that the fitting of data to the models of equation 4.15 does not come to an end with the fitting of the integral spectrum. As pointed out by Thiebaux and Dickie (1993), each of the domes, representing the oscillations in the overall distributions of density that give rise to the appearance of the secondary domes or trophic positions, can be analyzed according to the same principles used in analysis of the primary or integral slope. That is, if we recognize that predator-prey feeding relationships may exist within a trophic position as well as between positions, we have in the equation system a representation of the passage of food energy within this subsystem in terms of predator-prey relationships within domes, although perhaps exhibiting a smaller average body-size feeding ratio ($R$).

At this more detailed level, one might initially expect components of growth and mortality, to which much importance is attributed in the Platt-Denman (1978) continuous flow models, to have relatively greater effects on the biomass and production densities than when these characteristics are observed at the scale of the integral models. However, in our models of energy flow, changes due to these additional influences on average individual body size have already been accounted for by the sampling. That is, the fitting of the predator-prey models requires that we sample for the average biomass and production densities by body sizes for all of the organisms collected during a specified period. Plotting the average body sizes within a body-size category within a trophic position implies that we have already incorporated effects due to growth and mortality through the sampling design. These additional factors need not, then, be considered further in plotting the predator-prey biomass density profiles, which are, in the development of the theory, stable in time. We can thus turn directly to examining the secondary scale of predator-prey interactions.

If we apply to this secondary scaling the principle of similarity that we initially invoked in chapter 4 for the study of the primary predator-prey relations between trophic positions, we would expect regularities in predator-prey body sizes within the domes to be displayed as oscillations in biomass density within the trophic positions. That is, in parallel with the oscillatory behavior exhibited at the primary scale of fittings, we might expect to find a second set of oscillations in biomass density within the body-size ranges of these secondary domes.

Continued invocation of the principle of similarity, which we have used to examine the overall biomass spectrum, would also lead us to expect the following at the subsystem level of individual trophic positions: not only would we find the same type of oscillatory behavior as emerged in the overall-scale study, but there should be the same stability of relationships between the subdomes of biomass within trophic positions as between the original domes of the entire spectrum. That is, within the limits allowed by sampling vagaries, we would expect to find subdomes whose positions and spacing depend on the body-size ratios of the predators and prey within the trophic positions. These subdomes would appear along the locus of the biomass density profile of the trophic position dome. One might also expect, in a system that is linked by such important energy transfers as are represented in the predator-prey interactions, that the spacing of these secondary domes or oscillations would depend on the value of log $R$ that characterized the primary set of oscillations; that is, it would have a value that is either the same or is a simple harmonic of the initial value. These possibilities will be examined later in relation to the study of data.

In the case of the corresponding, secondary production spectrum, where the energy flow has a dominant effect on the biomass structure, it is also reasonable to expect to find higher slopes of production density over appropriate body-size ranges in the subdomes. The slope of the specific production spectrum depends on the value of $\gamma$, and the ratio of $\gamma$ to log $R$ determines the higher curvature of the secondary domes in the biomass spectrum. The possibility that a tertiary scaling might show up in the biomass density data was suggested by Dickie, Kerr, and Boudreau (1987) but the evidence supporting it was weak. Following the approach to analysis of data at the secondary scaling, Thiebaux and Dickie (1993) applied the biomass density spectral Model II to an analysis of the domelike distribution of biomass density that appeared in the data for the fish population of a small lake. Unfortunately, the evidence was again equivocal; the lake was quite small and contained no species larger than brook trout (*Salvelinus fontinalis*). Thus the data provided too few body-size categories to permit statistical verification of the possible two subdomes that were indicated in the population data. For this reason, Thiebaux and Dickie used the analysis of the small lake data only to illustrate and discuss the relative joint impacts of different predator-prey body-size ratios within food chains. Along with the effects of the conversion coefficients, these must affect interpretation of data fittings within the domes. We also reexamine these questions in the following sections.

No such limitations apply to the analysis of spectral data undertaken by Sprules and Goyke (1994). They demonstrated remarkable conformity of their empirical data to the expectations of Model II at both the primary and secondary levels of aggregation of data. We review their fittings here as the basis for judging the worth of pursuing these refinements of the models in relation to observing and managing natural systems.

## The Sprules and Goyke Fittings and Model Analysis

Data compiled from sampling cruises made at various seasons of the year in the two Laurentian Great Lakes, Michigan and Ontario, in 1987 and 1990, respectively, were analyzed by Sprules and associates. A comparison by Sprules and Goyke (1994) of the biomass density spectra for the two lakes is mainly based on the data subset that included phytoplankton, zooplankton, and fish. They note that the Lake Ontario sampling also included heterotrophic nannoflagellates and bacteria but that these groups were poorly represented in the earlier Lake Michigan samples. Their analyses of the data common to both lakes are especially notable in being based on diverse body sizes of organisms, deliberately sampled with necessarily diverse instrumentation in a manner that would yield a standardized annual average biomass density over the total designated body-size ranges. They have thus provided detailed information on the biomass density spectrum for the phytoplankton, zooplankton, and fish community of the pelagic zones of two major and previously well studied water bodies. Their results confirm that the elaborations of the simple theoretical expectations of the predator-prey models that we have been considering above clearly appear in the appropriate empirical fittings.

Sprules and Goyke (1994) initially present an unnormalized plot of the set of data from the two lakes. This enables us to glimpse the main characteristics of the two systems in generally familiar parameters. We have, therefore, copied their figure 2 in figure 5.2. It shows that there are three irregularly shaped domes of biomass density, corresponding with the phytoplankton, zooplankton, and fish of both lakes. These biomass densities are contained within generally flat integral density spectra, which differ in level between lakes by only a small amount. The regularity in the spectral densities of the two lakes becomes more evident in the normalized plots of the data, shown in figure 5.3.

Two features stand out in these results. The first, illustrated in the figures, is the coincidence of the body-size ranges of the three general functional body-size groupings in the two lakes. The second is the fact, pointed out by the authors, that the species represented in these groupings were quite different in the two lakes. Aside from the differences of phytoplankton species that are evidenced in the different peak densities

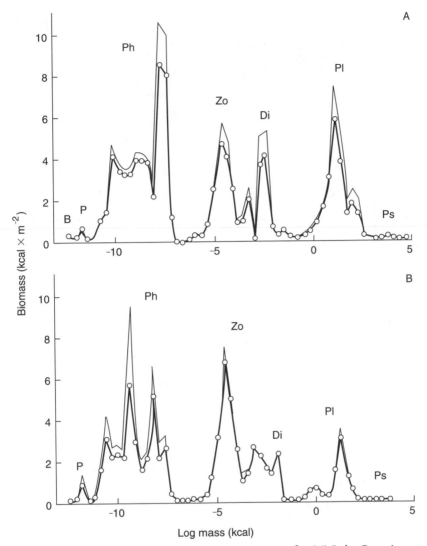

FIGURE 5.2 Mean unnormalized biomass size spectra for (A) Lake Ontario and (B) Lake Michigan. Circles and solid lines are means for individual size groups; dotted line is one standard error above means. B = bacteria; P = picoplankton; Ph = phytoplankton; Zo = zooplankton; Di = *Diporeia*; Pl = planktivorous fish; Ps = piscivorous fish. (Redrawn from Sprules and Goyke 1994.)

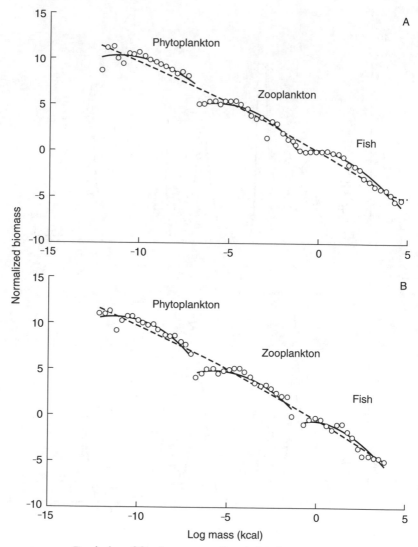

FIGURE 5.3   Parabolas of fixed curvature (lines) fitted to normalized biomass (circles) for individual trophic groups in (*A*) Lake Ontario and (*B*) Lake Michigan. Broken lines are parabolas fitted to the integral biomass spectrum. (Redrawn from Sprules and Goyke 1994.)

at different sizes in the two lakes (figure 5.2), in Lake Ontario the zooplankton consisted mostly of rotifers, bosminid cladocerans, and cyclopoid copepods; the smaller fish consisted primarily of the planktivorous alewife, *Alosa pseudoharengus* (79%), and smelt, *Osmerus mordax* (21%), which were associated with the larger fish—salmonids that have been introduced into all the lakes. In Lake Michigan, however, the zooplankton were dominated by *Daphnia* and calanoid copepods, and the small fish primarily by bloater, *Coregonus hoyi* (72%), which appeared in the fish dome together with the larger salmonids. That is, except for the salmonids, the similar primary and secondary features of the spectra that appear in the two lakes were generated by different species. These species have similar trophic functions in their respective ecosystems. The significance of this fact should not be overlooked.

The similarity between the lakes is made especially evident in the detailed normalized plots of biomass density within trophic positions for each of the two lakes. We show the results of Sprules and Goyke's fittings in figure 5.4. The theoretical expectation of Model II that similar parabolas will appear for the various trophically linked groupings is amply satisfied by the figures and the accompanying statistical tests. Residual variation around the generalized curves is greatest for the smaller phytoplankton sizes, a fact that the authors believe is partly a result of inadequate sampling. The rest of the curves show remarkable similarity in position and curvature, although the fish curve for Lake Ontario is noticeably more extensive and flatter, reflecting the higher densities of the large salmonids in this lake.

Of particular interest for our assessment of the reality of the model is the presence of apparently regular deviations of the biomass density from the fitted smooth, parabolic profiles of the trophic positions in both lakes. Sprules and Goyke (1994) thoroughly tested the suggestions of Thiebaux and Dickie (1993) for fitting data, and applied the fitting of the model to both the primary and secondary scalings—that is, for the integral spectrum and for the parabolic domes within trophic positions. They calculated deviations of the biomass density within the domes from the generalized fitted curves for each of them. We show their results in figure 5.5. In both lakes the deviations from the general curvatures of the trophic positions show up as a series of two or three organized, periodic peaks or subdomes within each trophic position. The

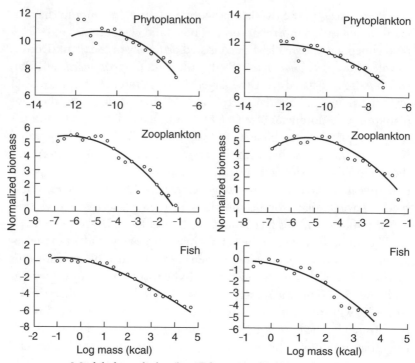

FIGURE 5.4   Modeled parabolas (lines) for normalized biomass density (circles) of trophic groups. Panels on left are for Lake Ontario, on right, Lake Michigan. (Redrawn from Sprules and Goyke 1994.)

correspondences among the subdomes for the two lakes are especially evident for the zooplankton and fish, but the patterns displayed by the phytoplankton are also similar, despite the different species represented throughout the two ecosystem series.

Sprules and Goyke (1994) statistically tested the fits of their empirical data to Model II. The results confirmed the evidence represented in the foregoing figures. Their calculations showed that the curvatures of the integral biomass density spectra for the two lakes had low values that were not significantly different from each other and were scarcely distinguishable from straight lines. The equivalent straight line slope values were very close to those published by Boudreau, Dickie, and Kerr (1991) and by Boudreau and Dickie (1992) for the integral spectral "lines" found in several marine ecosystems.

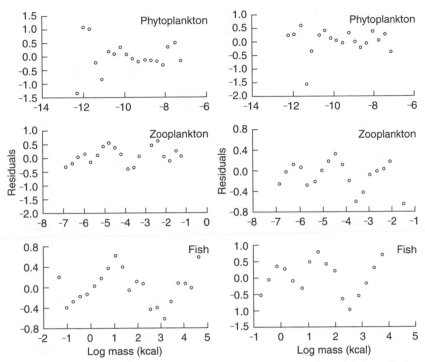

FIGURE 5.5    Residuals of normalized biomass density around modeled parabolas for trophic groups. Panels on left are for Lake Ontario, on right, Lake Michigan. (Redrawn from Sprules and Goyke 1994.)

With respect to fitting data within the various trophic positions, Sprules and Goyke's statistical studies, following the principles of similarity invoked by Thiebaux and Dickie (1992, 1993) sought, for each lake, the estimate of parabolic curvature that would give the best common fit of the three parabolas within each of the lakes. Their results confirm that the curvatures of the parabolas within lakes were not significantly different from one another, but that the set of parabolas for the more productive Lake Ontario had slightly less curvature than those for Lake Michigan.

More significantly, from the point of view of testing the general applicability of the model to the biomass data, they found that the vertical and horizontal shifts between parabolas within lakes had close to the same values. That is, their empirical findings indicate strong parallelism in the body-size ratios and the efficiencies of energy transfer between the predators and their prey at all trophic levels in both lake ecosystems. They also showed, as was implied by the Thiebaux and Dickie (1993) model, that the average body-size ratio between predator and prey within the trophic positions (between two and three $\log_2$ intervals) was measurably smaller than the corresponding average ratio between trophic positions (between five and six $\log_2$ intervals). Hence, not only were the parabolas similar within lakes, but the spacing of the domes and subdomes was similar in the lakes. Furthermore, the subdome horizontal spacing, representing the "within-dome" predator-prey body-size ratio, appeared likely to be a simple harmonic of the spacing between domes, preserving the principle of similarity that was invoked in the model building, and confirming that it repeatedly appears in the observed biomass spectral density patterns.

## Comparisons with Other Systems or Other Models

There are as yet no statistical studies of other data comparable to these whole-lake studies of Sprules and Goyke (1994) that would permit us to further test the fittings of the models to empirical data. However, Sprules and Goyke checked their results against what was available. For example, they compared the statistics of their empirical results for zooplankton in Lake Ontario and Lake Michigan to the parameters calculated from the spectrum compiled for the fish population of a small lake by Thiebaux and Dickie (1993). Sprules and Stockwell (1995) illustrated the applicability of the results in these two well-studied Great Lakes to the less complete information from Lake Erie, and both studies undertook to compare the production estimated from the spectrum for Lake Ontario with estimates obtained by more conventional methods.

The comparison by Sprules and Goyke of the dome derived from their zooplankton fittings with that of the fish population of the small lake studied by Thiebaux and Dickie provides a number of instructive contrasts. They found, for example, that the zooplankton curvature in Lake Ontario was only about one-half of the curvature for the fish population of the small lake. This is in keeping with the observations of Boudreau and Dickie (1992), who originally compiled the spectrum for the small lake, that it was so small that there seemed to be insufficient space in it to support fish species with large body sizes. Despite this environmental limitation, however, the small lake showed a coherent pattern of curvature in biomass density that resulted from the interaction of several species. That is, it showed characteristics similar to those found in the large lake samplings, with a curvature difference that was apparently related to the body-size range of predation opportunities.

The log $R$ ratio *within* the zooplankton dome of Lake Ontario appeared to have a value of $-2$, compared with $-3$ for the fish population of the small lake, a difference that may not be statistically distinguishable. However, it is notable that both values are only about half the predator-prey body-size ratio found between domes in both Sprules and Goyke's (1994) study, and by Boudreau and Dickie (1992) between trophic positions in their analyses of several major marine ecosystems. This appears to confirm the expectations of our earlier discussion that the subdome spacings may be harmonically related to the dominant spacing shown between trophic positions of the overall spectrum.

Sprules and Goyke's (1994) comparison is also instructive in relation to allometric values of the conversion coefficients, which we discuss more fully below. That is, in measuring the curvature of the trophic position domes for the Great Lakes, they adopted the terminology that we outlined in chapter 4. There the curvature of a biomass density dome, which we have defined as $c_0 = -\gamma/\log R$, takes a value for the allometric parameter $\gamma$, which may be considered as consisting of two parts: $\gamma' + \gamma''$. The first, $\gamma'$, depends on the allometry of metabolism; the second, $\gamma''$, depends on the postulated allometry of the conversion coefficients. Sprules and Goyke estimated the metabolic coefficient $\gamma' = -0.31$ for the zooplankton, which may be compared with Thiebaux and Dickie's estimate of $-0.29$ for the fish. This again meets the expectation of similarity. However, with large differences in the overall curvature of the subdomes, there are large differences in the estimate of $\gamma''$. That is, $\gamma'' = -1.05$ for the zooplankton but $-2.29$ for the fish. This is an unexpectedly large difference that needs to be evaluated in relation to factors that determine the allometry of conversion coefficients. We return to its interpretation in chapter 6.

Before leaving these fittings, it is useful to consider Sprules and Goyke's (1994) comparisons of production estimates calculated from the biomass spectrum with those calculated according to more conven-

tional methods. Platt, Lewis, and Geider (1984) made similar comparisons for estimates of production by the phytoplankton, using their linear model to provide revised measures. They considered the resulting
differences to be useful for detecting weaknesses in the more conventional methods of estimation. The comparison by Sprules and Goyke
(1994) uses information from the previous extensive studies of the zooplankton in Lake Ontario.

From the spectrum, Sprules and Goyke calculated the rate of zooplankton production, using allometric estimates for each body-size
group, summed over the observed total range. The results give an estimate of zooplankton production of 234 kcal $\times$ m$^{-2}$ $\times$ yr$^{-1}$, an estimate that would include production by all the species sampled, including
rotifers, epilimnetic and hypolimnetic crustaceans, *Mysis,* and *Diporeia.*
For comparison, direct estimates of production in epilimnetic crusta-
ceans alone have varied from 94.6 to 150.2 kcal $\times$ m$^{-2}$ $\times$ yr$^{-1}$ in different years, and a direct estimate of total production in another year, including the important *Mysis* and *Diporeia,* was 153.8 kcal $\times$ m$^{-2}$ $\times$ yr$^{-1}$.
Considering the practical difficulties that were experienced in making
the conventional estimates, which had up to that time excluded rotifers
and hypolimnetic copepods, Sprules and Goyke consider their higher
spectral estimate as the more realistic.

In another analysis, Sprules and Goyke (1994) compared estimates of production in Lake Ontario made by Borgmann (1983) using
Model I with their estimates using Model II. Borgmann's production estimate required the fitting of another parameter, a utilization coefficient
(different from, but related to what we have called the conversion coefficients), and also required an estimate of primary production in the
system. Even Borgmann's resulting low value of zooplankton produc-
tion—between 75 and 125 kcal $\times$ m$^{-2}$ $\times$ yr$^{-1}$, estimated according to
Model I—was considered by Sprules and Goyke to be an overestimate
for his methodology because of problems with the measure of primary
production that was available for use. The fact that Borgmann's estimate
was lower than production estimates that failed to take into account all
the known zooplankton indicated that Model I did not adequately represent the Lake Ontario data. The cause of the problem, which may arise
from either the fitting of the conversion coefficients or in the linear approximation to the biomass density shape, could not be assigned on the
basis of the available information.

In summary, it can be concluded that Sprules and Goyke's (1994) analysis confirms that the predator-prey model of energy transmission developed by Thiebaux and Dickie (1992, 1993) as Model II provides an adequate description of the biomass density distribution of organisms found in the pelagic zones of both Lakes Ontario and Michigan. The detail comprehended in the fitting implies that the goodness of fit depends on the degree of correspondence between mechanisms envisioned in the model and actual organismic predator-prey functions underlying the biological community; it did not depend on the particular species that occurred in the communities.

Notable, however, is the fact that evidence for the dominant importance of predation functions underlying the production functions of the Great Lakes systems appeared at the two levels of application: the integral fitting that described the system as a whole, at least from the smallest phytoplankton to the largest fish, and the fitting of subdomes that described the major trophic groupings within the system — the phytoplankton, zooplankton, and so on.

In addition to these conclusions, the Sprules and Goyke (1994) application carries important practical implications. In the first place, their analysis shows an orderly organization of the energy carriers at three levels of biomass density aggregation. The first level, that of the integral systems in each of the lakes, showed a measurable coherence, the common parameters of which could be measured as the position and curvature of similar overall parabolas. Within this integral fitting the secondary parabolas, the trophic positions, also showed strong similarity, both within and between lakes. The characteristic parabolic shapes at this secondary level could also be measured by a common value for curvature that reappeared at regular and similar body-size intervals in the two systems.

In addition, and remarkably, the analysis by Sprules and Goyke took the additional step of demonstrating the presence, within the well-known trophic positions, of a third series of subdomes or parabolas, describing a realistic, and apparently stable, inner pattern of predation relations, such as is implicit in the development of the realistic version of Model II. This is strong evidence supporting the speculation of Dickie, Kerr, and Boudreau (1987) that a tertiary level of trophic organization may be present in the biomass density of the living aquatic community. The original suggestion was made in connection with early attempts to

understand how individual species might play a definite role within ecological communities. The findings of Sprules and Goyke (1994) imply
that this tertiary level phenomenon does not comprehend all of the
characteristics of what we generally think of as a species, since different
species can be seen to play similar functional roles within the whole.
However, it is equally evident that the functions of species as ecological
entities in the trophic community are most clearly expressed at this tertiary level. It is an understanding to which we will need to return in
chapters 8 and 9, as we approach the practical question of how to manage fisheries for particular species in the context of the community production revealed by the spectrum.

## The Predator-Prey Body-Size Feeding Ratio—$R$

The foregoing review of the biomass density spectral properties suggests
that several features of the energy transmission invoked in the mathematical formulation of the underlying dynamics are observable in natural systems. Measuring parameters associated with them is critical to calculating production in the modeled system in relation to the biomass
density distribution. In particular, it has been observed that the passage
of food energy between predators and their prey takes place at two distinctly different body-size scales (i.e., between the three levels of organization of the biomass density). These seem to be characterized by two
different average values of the predator-prey body-size ratio—$R$. The
foregoing fittings, especially the detailed analyses of Sprules and Goyke,
confirm these basic patterns in the energy transmission. But in addition
to these empirical findings, the theoretical development suggests that
these two scales of body-size feeding linkage are likely to be related, re-
inforcing the expression of the principle of similarity invoked in the
original model developed by Thiebaux and Dickie (1992, 1993). In this
respect, the application of the basic predator-prey model at the two
scales of interaction would meet the most important criteria of reality
and completeness in the energy transmission system.

Predator-prey energy flow linkages within ecosystems must be the
principal mechanisms underlying the biomass density spectral structure
because food energy is the principal source of energy that follows the
primary production processes. Its expression will be modified by growth
and mortality patterns over the age distribution of a population; but

even for a situation where exploited fish populations are terminal natural predators in an ecosystem, calculations show that predatory energy transfers still represent by far the main pathways of transfer (cf. Bax 1991). That is, as stated at the beginning of this chapter, we are considering a model that appears to meet the criteria of reality required for both historical interpretation and predictive projection of ecosystem changes. However, while the mathematics of the expression of the parameters of energy transfer describes the basic form, observed perturbations originate from factors outside the mathematics. Therefore, we need to consider here the theoretical effects of variations in the values of the parameters used to explain the dominant patterns of spectral structure. Our purpose in what follows in this section is to consider the nature and the consequences of changes in $R$, the average body-size feeding ratio, on the spectral profiles at interaction sites at both physiological and ecological scales.

## Variations in Feeding Links Within a Trophic Position

Perhaps one of the most striking aspects of the solutions for the spectral equations of Model II in equation 4.15 is that, apart from the evident periodicity or repetition of the parabolic domes that is implied in the equation system, the mathematics alone can really say little more than that the periodicity must depend on the body-size feeding ratio. That is, the formulations of the prey-predator model demonstrate the existence of a periodicity that results from the feeding interactions and has very important implications for interpreting the distributions of the ecological elements. However, the details of this periodicity are determined by factors outside the model framework.

The mathematical solution implies, however, two important facts. The first is that the average value of the periodic structures, which we have called $J_0$, if it exists, can be absorbed into the general measure of the level of the system biomass density, $B_0$. In a sense, it may be regarded as a free constant, the value of which is unlikely to have practical meaning separate from that attached to interpretation of the general level or intercept of the spectrum. This was discussed in chapter 2 in relation to the integral spectrum and so need not detain us further here.

The second fact concerns the interpretation of the oscillatory part of the solution that we have termed $J(\log w)$. It represents the residual oscillation of the biomass density about the basic quadratic part of $\log B$, with a period related to $|\log R|$. These oscillations of the biomass with body size within trophic positions would be stable in a simple model where the average feeding body-size ratio has no variance. In fact, the low variance of its value implied by the Sprules and Goyke (1994) fittings suggested a goodness of fit that far exceeded our a priori expectations. The secondary domes revealed within trophic positions in their study therefore invite further explorations of their characteristics and role in the dynamics of natural ecosystems.

In situations where there may be a higher variance of the value of $R$, such as may well occur with an increase in the range of body sizes of food items selected by a predator, a negative feedback mechanism would likely act to dampen the oscillations. For example, in a real system, if a predator on the crest of an oscillation feeds on prey particles that are near but not precisely on the crest that describes the prey biomass density, it would be feeding on a prey biomass density that is lower than that which the simple model calls for if the overall biomass density distribution profile is to be maintained. That is, departures from the precise body-size ratio of a predator-prey pair would call for higher prey densities than the simple system would afford at these additional body sizes. The continued existence of the oscillation in the presence of such departures would therefore depend on possibly complex interactions of predator foraging behaviors and prey availability. But oscillations cannot for that reason alone be dismissed as possibilities. In keeping with this speculation, it is worth noting that the size range of food items selected by predators seems to be widest at times of temporary increases in food abundance; in times of prey scarcity, predators tend to become more narrowly specialized (Zaret and Rand 1971; Keast 1985). The significance of such interactions in natural ecosystems needs to be studied in further detailed investigations of predator-prey interactions under various conditions; their persistent occurrence seems to be consistent with the stringent requirements needed to maintain predator-prey density balances.

In consequence, variance in $R$ about its average values would be expected to produce a smoother spectrum than the simple model predicts. We have too little experience with such measures to say much more about them, except that the apparent sharpness of the subdomes that appeared in the annual average data compiled and analyzed by Sprules and Goyke (1994) provides reassurance that the average value of $\log R$

is a phenomenon with a detectable existence that should be amenable to study.

The Sprules and Goyke (1994) study helps us emphasize another point about interpreting basic trophic interaction phenomena. That is, the oscillations we are discussing with respect to body size are *not* oscillations in time. This was demonstrated empirically by the studies of Vezina (1986) and Rodríguez and associates (1987) that we reviewed in chapter 3. The point is made again here by the fact that the Sprules and Goyke fittings are based on annual averages of data, collected over significant periods within the production seasons in these temperate zone lakes: the domes and subdomes both appeared stable in time.

We need to recognize, nevertheless, that equilibrium values for real systems will be affected by time-dependent factors at possibly all three of the different scales at which we have detected persistent structural features. That is, the equilibria implied by persistent structural features must arise from interactions that take place on quite separate orders of time relations (Denman, Freeland, and Mackas 1989). The overall equilibrium of the distribution of organism body sizes within a whole ecosystem, what we have called the physiological scaling of the biomass density, must be dependent on many factors that in the past have supported the metabolic adaptations of the body sizes of organisms making up the existing communities (cf. Warwick 1984). This must include characteristics that are reflected in life histories of the individual species, including the size at maturity, fecundity schedules, and the sizes at which, as predators, they can capture different sizes of prey. These are all mechanisms through which the body-size distribution of the whole community is expressed and must depend on evolutionary time scales.

While changes in the expression of physiological characteristics in an individual species must depend on relatively long time scales, the interactions through which they are expressed within the community must be renewed seasonally, in keeping with their temporally varying environments. We should therefore expect that, given a certain fauna, biomass density profiles may still exhibit the effects of year-to-year changes in system fertility. That is, detecting spectral phenomena at the integral spectrum level seems likely to depend more on developing sampling schedules that provide reliable annual averages than it depends on development of long-term equilibrium conditions in the individual component populations. Spectral data may well be profitably examined over longer time periods, but this requirement is more related to variations in the distributions of species in their geographical ranges than to failures of the biomass density to meet theoretical equilibrium conditions of the component species.

By contrast with the longer time scales related to the expression of physiological factors in individual species, the secondary or ecological scale of the predator-prey relationships—on which the internal structure of the spectrum depends, and on the basis of which the predator-prey model was formulated—must affect biomass density distributions on relatively short time scales. We would expect them to respond to changes in the ecosystem on the order of hours to days, or at the most weeks. This is the order of time that is required for a predator aggregation to adjust its density to a level where the metabolic requirements of individual predators will be satisfied by the intake of food afforded by the existing prey density (Hochachka and Somero 1971). One might therefore expect that densities of predators and their prey in spectral profiles would need to be sampled within specified limits of accuracy. For phenomena operating on short time scales, these should be afforded by the usual space and time considerations for estimating present population density.

There remains a third level of time variation that is evident in the contrast between species representations in Lake Ontario and Lake Michigan. The analysis of Sprules and Goyke (1994) showed that, while secondary structures reflecting predator-prey body sizes showed up as stable features of the biomass distributions, they were made up of quite different species. Species equilibria within communities present us with important practical questions, which at present have too few answers. We will return in chapters 7 and 8 to the question of how variations at this scale might be taken into account in connection with highly important management concerns. For present considerations, however, we conclude that variations in the value of the average feeding body-size ratio will be reflected in the curvature of the fittings of the biomass density in both the trophic position domes and subdomes. Hence, sampling for the spectrum needs to be directed primarily at density within body-size classes; and only for reasons outside the fitting of spectral parameters does sampling need to be directed toward separating and identifying the species involved.

## Variations in Feeding Links Between Trophic Positions

In addition, it is necessary to consider the effects of variations in the average values of log $R$ that may be observed in different parts of the ecosystem, such as between zooplankton predators or fish predators in relation to their respective preferred prey in other trophic priorities. To inquire about the effects of variation at this level of aggregation, essentially on the integral spectrum, we again recall the allometric form that we have assumed for the feeding body-size ratio. In chapter 4 we defined this ratio as

$$R_j(w_j) = w_i/w_j = Rw_j^p,$$

where in model development to this point we have assumed that $R_j$ is a constant. In the foregoing terminology this is the equivalent of treating the exponent p of $w_j^p$ as equal to zero.

We can examine the general effects of changes in the feeding size ratio between trophic positions by letting p take values greater than zero. To begin, we return to the designation of predator and prey groups by the labels $j$ and $i$, respectively. Recalling the formulation of Model II in equation 4.19, we wrote the difference equation

$$y_j(x) - y_i(x + \log R) = \log \alpha + \gamma x,$$

which exemplifies the central place of log $R$ in the system. In the original biomass density terminology, this may be written as

$$\log B_j(w) - \log B_i(Rw) = \log \alpha + \gamma \log w. \qquad (5.1)$$

This equation has multiple solutions, in keeping with the multiple-valued case we studied in chapter 4. That is, it has solutions labeled $i$, $j, \dots$, which would in fact represent independent distributions of the biomass density of groups $i$ and $j$, representing prey and their predators, were it not for the trophic linkage between them. But the principle of similarity that we have employed throughout makes it evident that if the biomass spectral density of one feeding group is arbitrarily specified, then the spectral density of the other feeding groups linked with it must also be predicted by equation 5.1. That is, with variations in the body-size feeding ratio we are evaluating the strength of the effects of this linkage on the observed biomass density.

Let us suppose that we take the predator spectral density to be a parabolic dome with a curvature $c$ located at the position $(\log w_{j0}, \log B_{j0})$. Then we may write

$$\log B_j(w) = \log B_{j0} + \tfrac{1}{2}c_0[\log w/w_{j0}]^2. \qquad (5.2)$$

Equation 5.1 predicts that the corresponding prey spectral density is the parabolic dome

$$\log B_i(w) = \log Bw_{i0} + \tfrac{1}{2}c_0[\log w/w_{i0}]^2. \qquad (5.2a)$$

That is, the prey spectral density has the same curvature as the predator dome, but its vertex is located horizontally at

$$\log w_{i0} = \log w_{j0} - h$$

where

$$h = -(\gamma/c) - \log R$$

and vertically at

$$\log B_{i0} = \log B_{j0} - [\gamma^2/2c + \gamma \log w_{j0} + \log \alpha].$$

While it is clear in this formulation that the horizontal steps in the sequence are functions of $\log R$, we note that the vertical steps in the sequence are independent of it. That is, if the biomass spectrum of the initial feeding group of predators is a parabolic dome, then the predicted biomass spectrum of the prey is also a parabolic dome having the same curvature, but displaced vertically by an amount related to the metabolic coefficients of individual body size, and horizontally by an amount that is related to both the metabolic rates and the predator's feeding body-size ratio.

Our objective now is to examine the effects of replacing in equation 5.1 the constant $Rw$, which we now understand as $Rw_i^0$, with an expression that we may write as $Rw_i^{p+1}$, where the exponent for the prey group, which we may designate as group p, is written as p + 1, which can now take a range of nonzero values.

For equation 5.1 in terms of the biomass density of the new set of prey, we may now write

$$\log B_j(w) - \log B_p(Rw^{p+1}) = \log \alpha + \gamma \log w. \qquad (5.3)$$

In parallel with the partial solution of equation 5.1 given in equation 5.2, we find that the new prey spectral density predicted by our system is the parabolic dome

$$\log B_p(w) = \log B_{j0} + \tfrac{1}{2}c_p[\log w/w_{p0}]^2.$$

In the case where the body-size selection of prey by the predator has a different average value than that which characterized its own biomass density dome, the predicted prey density will have a different curvature:

$$c_p = c/(p + 1)^2.$$

The horizontal position is given by

$$w_{p0} = \log R + (p + 1)(\log w_{j0} + \gamma/c)$$

where it is to be noted that the horizontal distance between the predator and prey dome may be equal to, greater than, or less than the value of $\log R$, depending on the sign of $\gamma$. That is, the mathematics alone does not make an a priori statement about whether a predator at the peak of the predator dome feeds precisely at the peak of a prey dome. There are special cases of general interest: for example, the arctic is a special environment, which has highly pulsed production systems with few highly specialized predator-prey relations. It can be treated in our terminology by variations in the value of p, which in the case of very little range in the choice of prey sizes will tend toward a value of $-1$.

In such a case the curvature of the prey dome approaches $-\infty$, the dome itself appearing as a narrow vertical line at the fixed prey size $w_{p0} = R$. A parallel situation could appear in the case of a predator that had a very highly preferred prey species. Considering the many behavioral and morphological adaptations that exist in natural communities, and the many metabolic niches that have been filled by appropriate predator-prey systems, a very broad range of situations may eventually arise in detailed studies of ecosystems. It should be possible to treat most of them in the terms introduced here.

It is notable in relation to these derivations that the vertical shift in the position of the predicted prey dome is again independent of the value of the feeding body-size ratio, hence of the value of p. In parallel with the previous derivations, its vertical position is given by

$$\log B_{p0} = \log B_{j0} - [\gamma^2/2c + \gamma \log w_{j0} + \log \alpha].$$

To this point we have attempted to fit the parabolas using only information on the relative body sizes of predators and their prey. While this permits us to perceive the effects of different predator-prey interactions that may be reflected in the statistics of the domes, remember that we have still to deal with the effects of differences in production efficiencies. Since they are also treated as allometric functions, their values will further modify the spacing and shapes of the domes. We discuss their effects in more detail in the following chapter.

## Summary and Some Implications

We have examined the specific effects on the model output of relaxing the requirement for a constant predator-prey body-size ratio to describe the passage of food energy from one trophic position to another. This is accomplished by applying the same principles utilized to analyze the integral spectrum to study the body-size intervals within the trophic positions themselves. The theoretical developments of this step seem to be verified in remarkable fashion by the empirical fittings demonstrated by Sprules and Goyke's (1994) study of data for two of the Laurentian Great Lakes. Their analysis shows that not only do the trophic position domes of the biomass spectrum show strong similarity throughout the successive stages in the predator-prey food chain links, but that within these domes there appear to be appreciable and stable secondary domes that represent the expected food-chain linkages within trophic positions.

Theory suggests, and the data seem to verify, that these internal linkages are probably simple harmonics of the size-interval relations established in the integral spectrum between the trophic positions themselves. It is notable, however, that Sprules and Goyke's analyses show that variations between ecosystems may well be reflected in the curvatures of the trophic position domes. These curvatures were different for the two major lake ecosystems, although a common curvature appeared to be maintained among the trophic domes within a single ecosystem.

We have studied the effects on the parameters of the spectrum of variations in the body-size ratio of predator-prey feeding pairs, and concluded that a change in the body-size range of prey taken by a predator will, indeed, lead to a change in the curvature of the parabolic dome describing the prey distribution. Because prey and predator distributions are linked through the feeding process, this appears to lead to changes in the predator dome as well, with commensurable, though not necessarily small, effects on the slope of the integral spectrum. In particular, it appears that variations in the predator-prey body-size ratio would lead to the development of a smoother spectrum than that expected for the theoretically simple case of a stable, single pattern of prey body-size selection. Evolution of the terminology describing the parameters of the feeding body-size ratio allows the examination of special ecological situations, such as where predators have a very limited size range of prey from which to choose.

The persistence of parabolic domes throughout these general cases of change in the feeding body-size ratio is notable. It underscores within the whole development of this model the special status of parabolic domes. That is, parabolic structures seem likely to be special features in the description of the density distributions of biomass by size classes because ecosystems seem to be dominated by predator-prey energy linkages. It should be noted that, mathematically speaking, parabolas are the only shapes that would be replicated throughout the system in this fashion without distortion. Parabolas thus accord with the similarity rules that we have embodied in equation 5.1, and that appear from the data to be operating across feeding groups in the whole ecosystem. It is tempting to speculate that the parabolic form may have a special status in real ecosystems. Before concluding further about such fittings, however, it is essential to take into account the influence of the conversion coefficients.

We have now set out the fundamental structure of our line of reasoning. In the remaining text we examine some of the essential underlying processes, and then explore their consequences.

∾

# CHAPTER 6

# Physiological Bases for Spectra

"Explanation" of observed behavior of a complex system can be sought at various levels of organization. In this sense, explanation is primarily a function of the interests of the analyst. At one level of analysis, for example, we have already provided a robust explanation for the existence of orderly biomass density spectra in aquatic communities; their characteristic form is an inexorable consequence of mathematical relations between size-dependent allometries of the acquisition and transmission of energy through predation processes and the attendant metabolic dissipation. Allometry is itself a statistical commonplace in physiology, sufficient that some readers may be content with explanation phrased at that level of aggregation of observations. Others will not, especially those interested in energy flow mechanisms, with the purpose of understanding the specific nature of the controls on production that act through them. For that reason, we need to turn now to an additional physiological level of explanation that depends on compilations of observations at still more detailed degrees of aggregation.

In this chapter, while we are primarily concerned with the level of integration represented by physiological processes, our interest is not so much in the processes themselves as in how they provide the means through which the emergent properties of system production become distributed in particular ecological patterns through the aquatic ecosystem. For that reason we are particularly concerned with the physiological factors underlying the conversion factors, $K$ and $F$, identified in chapters 4 and 5. These are the ecological efficiency and predation efficiency parameters, respectively, encompassed by the Thiebaux-Dickie model. Physiological properties underlying $K$ and $F$ arise among individuals, but their effects emerge at the level of the integral spectrum as well as in connection with the internal structure that can appear as various biomass domes and subdomes within it. Limits to interaction set by physiology in the context of system production thus contribute to understanding how emergent behaviors are conserved in the system. This level of explanation assists us, in subsequent chapters, to understand the higher levels of integration represented in the lines of control that relate to society's interests in sustained exploitation of aquatic natural resources.

## Physiological Processes as Mechanisms

Through physiology we are concerned with processes that are manifested and measurable at a hierarchical level of organization that is lower than the resulting ecological production processes of the community. Explanation at these lower levels of organization is conventionally thought to provide a closer link to the physics and chemistry regarded as embodying the truly "scientific" phenomena of quantitative measurement and prediction. However, the view of system behavior that emerges from physiological investigations can support a much wider view of the causal nature of dynamic biological systems than is afforded by reference to the underlying physics and chemistry. Physiological processes reveal underlying mechanisms, many of them nonlinear, but they are also essential elements in the formation of integrated feedback informational responses at higher system levels of interaction. It is this organizational aspect of physiology that is of most direct interest in production studies.

We have also already noted that mechanisms operating at the physiological level are *primary* in a different, quite specific measurement sense: they appear to lie at the base of the integral structural form of the biomass spectrum. That is, they play a decisive role in determining the values of the overall slope or curvature of both production and biomass density spectra and of their intercepts. In our equation systems, therefore, mechanisms operating at the physiological level express the interactions of organisms as these appear at the primary level of scaling of the efficiency coefficients introduced in equation 4.6. We specified them as the energy conversion and predation parameters, $K$ and $F$, and part of our concern is with assessing the effects of changes in their values at both primary and secondary levels. As indicated in chapter 4, expression of these values as allometric functions of body size at both physiological and ecological scales considerably simplifies mathematical expression of the spectra.

Appeal to underlying process, however, is premature unless the emergent behavior of the system for which explanation is sought has been clearly identified first. To choose a somewhat archaic illustration, the purpose of the mechanical clock is to index the passage of time by rotating its main shaft at a fixed rate that is independent, within broad limits, of the power available to the system (i.e., the tension of the main spring). The critical component regulating this behavior is the escapement mechanism, which is typically an inconspicuous element amid the complex assembly of gears and springs that compose the major parts of the mechanism. Accordingly, the phenomenological behavior of the clock, as a device to index the passage of time, must first be understood before the analyst can begin to place in perspective the importance of internal details. The escapement mechanism is crucial. All the remainder of the mechanism, complex though it may seem and as prominent as it is, is merely a power train with gear reductions to allow the escapement to index the earth's rate of rotation through the movement of the main shaft.

The metaphor of the clock has been overused in the biological literature. Our comments apply with equal force to mechanical devices and modern timepieces (atomic, electronic, etc.), which are, without exception, essentially oscillators. The once familiar "tick, tock" of the mechanical escapement is merely replaced in modern times by the sine wave oscilloscope trace of an electronic timepiece. The principles remain the same, though digital chips do not afford the same tactile delight as the mechanical devices of an earlier generation. The point is that any such oscillatory behavior should alert the analyst, at the emergent level of analysis, to the importance of periodic entrainment in the behavior of the system.

We reintroduce the mechanical metaphor quite deliberately here, despite the fact that younger readers may have no direct experience of such technical anachronisms, because the clockwork metaphor has been used, and profoundly misunderstood, by otherwise able ecologists and others, for a very long time. To our knowledge, use of the clockwork as biological metaphor dates at least to Paley (1802), and its usage has not improved with time. The mechanical analogy fails to recognize the critically important information and energy content of the organization patterns. Ecosystem adaptability at the higher levels depends on the feedbacks and flexibility in response at lower levels.

Throughout what follows, we emphasize the view that to explain a complex system is first to apprehend its emergent behavior, and secondarily to show how that behavior is maintained and manifested in structure. It is not enough to reduce the system to a heap of gears and springs: countless children of the authors' generation (the authors included) have amply demonstrated the insufficiency of that a priori approach to reductionist analysis of mechanical clocks. As Rosen (1991) has pointed out, even at the level of distinguishing organism from machine, first must come the phenomenological description, and only then can effective internal analysis follow. The metaphor of ecological systems as clockwork mechanisms may, therefore, be seriously misleading, and in any case is not otherwise very compelling. We use it only to establish the important point that small children, and others, can learn about the salient properties of complex systems only by first observing their phenomenological behavior.

## Organization of the Production System

Production is an ecological process, and fish production reflects the influence of a variety of internal and external factors operating over a wide range of spatial and temporal scales. In this section, we consider how the production process is organized, what are the principal elements in the hierarchical nest of external and community factors within which the internal factors are embedded, and how internal factors are expressed in physiological terms.

Figure 6.1 broadly outlines the approach. In some sense, it is an aggregate view of the algebraic formulation set out in figure 4.1, but phrased here at the level of a self-reproducing stock of organisms. The schematic depicts, in this instance for fishes in particular, the familiar progression of morphogenetic development from eggs through to sexually mature adults. An integral aquatic production system comprises an intricate network of such patterns. At the outset, this much alone entails recognizing a significant difference between the ecology of fish produc-

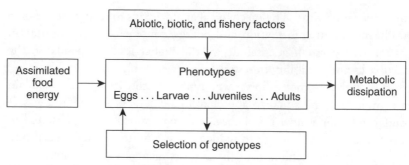

FIGURE 6.1    Interaction of the energy and information transfer systems that determine biomass production. The horizontal arrows represent energy flow through phenotypes. The vertical arrows represent information flow from external factors, acting upon phenotypes to condition production, which may in turn select the genotypes that determine the production potentials of future generations.

tion systems and that of the more familiar terrestrial endotherms, such as the birds and mammals.

The scheme set out in figure 6.1 reminds us that the large-bodied fishes, like many other organisms, typically undergo a remarkable series of transitions in their utilization of the food chain as their growth and morphogenesis progress. This is depicted by the horizontal arrows, representing the energy flows that support the process. In this respect, fishes resemble terrestrial insects, where the trophic habits of larvae are often very different from those of the adults. This characteristic trophic progression is quite unlike that of most birds and mammals but applies broadly among aquatic heterotherms.

The effects of other environmental factors must also be contemplated. This brings us to the upper box in figure 6.1, where the various possible environmental effects that can modify the production process at any stage of morphogenesis have been separated into abiotic factors (e.g., temperature, oxygen tension and concentration, and salinity), biotic factors (e.g., prey availability, parasitism and disease, competition, and predation), and the special and important effects directly stemming from fishery predation (e.g., size-selective mortality).

We use the term "factor" in the sense offered by Fry (1947, 1971), as an "effector" that causes changes in the organism's response to its environment. Our purpose in the balance of this chapter is to consider the capacities of these factors to effect changes in aquatic production. We also wish to emphasize the range of temporal and spatial scales over which these effects are characteristically observable. We should note that our separation of external factors into discrete categories is a matter of convenience for the purpose of explication; it does not imply that they operate independently.

Phenotypes can themselves exert numerous important effects on the environmental factors that affect them, particularly on the associated biota and even on the physical attributes of the environment. But for simplicity, we have not attempted to depict these feedback effects in figure 6.1, although we discuss some of them at length in subsequent chapters. The figure also fails to capture any sense of the importance of spatial variation, another subject we deal with subsequently.

Finally, it is important for this study to note the prominent genetic feedback loop in figure 6.1, depicting the selection of successful genotypes for inclusion in the genetic pool of the next generation. This effect is summarized by the vertical arrows shown in the lower part of figure 6.1. Genetic transmission of information is often treated as separate from the transmission of energy; certainly the tracing of its effects requires analysis according to its own specific set of rules. However, we cannot neglect the remarkable relations between the genetic and energetic sources of maintenance, nor the differences that may distinguish their interactions in aquatic organization systems from those of their terrestrial counterparts.

For example, a significant contrast between the physiology of fishes and terrestrial homeotherms that strongly affects their ecology is the relatively much greater fecundity of fishes, which is apparently related to adaptation, during their long history, to an environment very different from the terrestrial. The effects, of course, also differ in magnitude among the various taxa within these two major groupings, but of most consequence is the difference in pattern between the two major sets of responses to their environmental settings.

Among aquatic vertebrates the production of gametes can be extraordinarily generous, relative to that of terrestrial forms. A large female cod, to choose an extreme example, will produce literally millions of eggs annually, when only two, on average, must survive over its whole lifetime of 20 or more years of reproduction to replenish the stock. The unfished cod stocks of earlier centuries supported fecund females to even greater ages, giving rise to a population expenditure of energy in the exercise of fecundity that is quite beyond the experience or even imagination of terrestrial vertebrates. Indeed, it is interesting to speculate that cod larvae at one time contributed substantially to the plankton and thus to the food chain, supporting both larval and postlarval forms of many species. The attendant selection pressures experienced by a cod stock, whether exploited or not, must be many orders of magnitude more intense, and distributed over the stock elements in a quite different manner than would be experienced, for example, by a stock of caribou, or gray seals. For reasons of this sort, one cannot be content with explanations of the organization of aquatic communities that overlook the genetic feedback system that operates in conjunction with the physiological processes. We examine some of the consequences of these ideas in more detail in chapter 8.

It must also be clear in this line of analysis that the aquatic heterotherms (poikilotherms), both vertebrate and invertebrate, pursue different strategies for survival than the more familiar terrestrial homeotherms, the birds and mammals. There is, however, the large group of aquatic homeotherms, the seals, dolphins, and whales, perhaps including even the seabirds, for which the spectral information is still so fragmentary that we are not yet in a position to deal with them as we do other aquatic organisms. Nevertheless, we have no reason to suspect that their special adaptations to life in the marine environment make their population structures or production characteristics different from what can so far be seen for the aquatic heterotherm groups. Apollonio (in preparation), in a subsequent book in this series, assembles estimates of the former and current abundances of mammals and seabirds in the Gulf of Maine. His estimates, while unavoidably rough, are consistent with our expectation.


## The Energetic Basis of Aquatic Production

All biological, energy-flow production systems possess certain physico-chemical characteristics in common. That is, whatever the spatial and temporal scales of system definition, each is subject to the same constraint of energy balance. The energy balance equation requires that input to the system be channeled along one of two alternate routes: either the assimilated energy is retained as new biomass, or it is dissipated as metabolic expenditure. If we put aside for the moment the question of energy transfers across system boundaries, the foregoing truism is a logical requisite whether the system under consideration is a temporary forest pond or the Gulf of Thailand. It holds with equal force whether the system embodies an individual fish in an aquarium, or the entire biota of the Scotian Shelf.

This initial definition reflects nothing more than the simple fact that aquatic production, regardless of its scope or provenance, is inescapably a reflection of the thermodynamics of the open thermodynamic system of which it is a part. From this simple premise a number of fascinating consequences flow, some of which we need to examine here.

## The Energy Balance Equation

This logical prerequisite has been set out in various forms. The formalism we use, together with the attendant symbols, is unavoidably pragmatic owing to our need to avoid confusion with similar terms used in the earlier literature. We write it as

$$pC = T + S, \qquad (6.1)$$

where $C$ is the ration intake or consumption, corrected by the dimensionless coefficient, p, for nonassimilation of indigestible content and waste products (see figure 4.1), $T$ is the total metabolic dissipation, and $S$ is the energy retained as body substance or mass, all measured over the common interval of time, $\Delta t$, in conventional energy units: calories or joules. The $S$ component may take the form of either somatic or gonad elaboration, or both. The general term offered for this elaborated biomass component by Ware (1980) is "surplus energy." Ware's intent was to indicate that surplus energy is the currency, above and beyond its immediate needs for survival, that an organism can "choose" to divert into either somatic growth, or reproduction.

Initially formulated to describe the bioenergetic performances of individual fishes, the truism phrased by equation 6.1 can be invoked at various levels of stock, population, species, or species groups within communities. A great deal is known at various levels about some of the details of the terms in equation 6.1—how they vary in response to external factors, for example, especially in relation to individual fishes—

but much remains to be explored. Here we emphasize two physiological levels within the system: the components of total metabolism and the functional form of the conversion coefficients as they are identified in chapters 4 and 5, both, of course, in relation to changes in body size that are the basis of the biomass spectrum. Initial interest is at the level of the physiology of individual organisms, but we necessarily expand the analysis to encompass the properties of population and community that give rise to the spectrum.

## The Metabolic Component

The metabolic component, $T$ in equation 6.1, reflects the portion of energy assimilated that is subsequently dissipated, ultimately as heat loss to the environment, in the process of supporting activities that devolve to the well-being of the organism. We consider some of the essential components of $T$ in a moment. For now, we merely point out the often-observed property that $T$ is an allometric function of body mass. The power index of the relationship varies according to the components of $T$, and the circumstances under which it is measured, but a value of $0.8-0.85$ is common for fishes and many invertebrates (Fry 1971). This mass-dependent allometric scaling has been referred to in previous sections as causing the "physiological" component of the biomass spectrum.

## The Growth Component

The surplus energy term, $S/\Delta t$, representing the potential for growth in individual organism body size, offers the most immediate consequences for the formation of regularities in size composition in the biomass spectrum. The secondary scalings of biomass structure described in chapter 3 exist because of basic similarities in individual growth rates at various levels of the ecological hierarchy, their population variance as a function of body size, and the vicissitudes of natural mortality, all of which must have a decisive effect on the critical development of the predator-prey body-size ratio. We clarify this assertion in the following considerations.

While fish growth, like invertebrate growth, is often described as "indeterminate," this is something of an oxymoron. The growth trajectory

followed by individuals in most heterotherm groups is highly determined, hence predictable in principle. The salient difference between
"indeterminate" and "determinate" growth is that the former is responsive to a much broader range of external factors than is usual with the
strong, determinate, hormonal control of growth among homeotherms.
When birds or mammals fail to encounter conditions necessary for their
relatively restricted patterns of growth and development, they die. When
fishes or other heterotherms experience inimical conditions, they typically manifest a broader range of responses, which can in turn result in
an array of growth or developmental patterns. This is a heterotherm
strategy that is not available to homeotherms, and may go some distance
in explaining why the overwhelming majority of animal species have
elected to pursue the heterotherm strategy; it is both more flexible and
less energy demanding per unit biomass than the homeotherm alternative, even though, as pointed out by Humphreys (1979), the ecological
difference (i.e., production per unit area of habitat) may be far less
than appeared initially from the physiological measurement of isolated
individuals.

Fishes and other heterotherms have a wide variety of responses to
habitat variables. They are generally competent to accommodate to the
normal variation of environmental factors such as temperature. However, they are far less labile than homeotherms when encountering environmental variation beyond the norm. Thus, the heterotherm strategy
requires metabolic investment in the premise that environmental conditions will not deviate much from a particular, sustainable range of values. Homeotherms, on the other hand, reserve a capacity to respond to
changing circumstances. Perhaps it is natural that we, as analysts and as
homeotherms, tend to regard this "independence" strategy as somehow
more natural or advanced—even enlightened. The fact is, however,
that heterothermy is overwhelmingly the favored aquatic strategy, as indeed it is throughout the animal kingdom, implying that, ecologically,
homeothermy may be a metabolic luxury permitted to those few who
can afford it for special purposes.

To phrase things somewhat more generally, there is nothing in equation 6.1 to restrict its application to the fishes or other heterotherms; it
is merely a truism of nonequilibrium thermodynamics that applies with
equal force to any living organism. What is different for heterotherms
becomes evident only upon deeper analysis. Among other possibilities,

the approach we use here in relation to the biomass spectrum is to consider the efficiencies of energy use that can be described in relation to the conversion factors $K$ and $F$, of equation 4.6.

## Growth Efficiency—K

To show how $K$ is affected by environmental factors, we begin by examining $K$'s physiological basis.

### PHYSIOLOGY OF INDIVIDUALS

Transposing equation 6.1 to the form

$$S/pC = 1 - T/pC = K \tag{6.2}$$

provides a simple definition of the net growth efficiency of individual organisms as the fraction of the assimilated ration that is retained as body substance. In earlier chapters (figure 4.1) we have designated it by the symbol $K$. Equation 6.2 states that individual net growth efficiency, $K = S/pC$, scales inversely with the metabolic expenditure per unit of utilized rations.

We first turn to the effects on $K$ of the metabolic term $T$ in equation 6.2. $T$ is customarily regarded as composed of several components. Some components vary in response to physical environmental factors (e.g., temperature); others vary in response to biotic factors (e.g., food supply). And, as noted earlier, the metabolic rate itself scales allometrically with body mass. Accordingly, these various components are often expressed in the form

$$T = T_S + T_F + T_{AHI}. \tag{6.3}$$

In this formulation, $T_S$ is the standard metabolism, defined as the cost of maintaining life under specified environmental conditions and in the absence of other challenges. The two remaining terms figure importantly in relation to $C$: ration acquisition and its processing cost. The activity metabolism, $T_F$, encompasses the expenditures needed to support prey capture, among other activities. The Apparent Heat Increment, $T_{AHI}$, is often described as the SDA (Specific Dynamic Action), and can be thought of as the entropic tax on processing ingested food (Ware 1980).

To give some idea of the approximate magnitudes of these metabolic components, the standard metabolic cost of maintenance, $T_S$, typically varies from perhaps as much as one-fifth of the total aerobic limit of metabolism for a cod (Kerr 1982), to perhaps one-ninth of the aerobic limit for a sockeye salmon (Brett 1965). On the other hand, $T_F$ can range from relatively insignificant levels, as it is especially under most laboratory conditions, to a substantial portion of the total metabolic expenditure, especially under ambient circumstances in the field where a fish must search for and secure its ration (Kerr 1982; Krohn and Boisclair 1994).

In general, $T_{AHI}$ has been thought to embody perhaps 15% of the ingested energy. Following the additive model of equation 6.3, it has also been assumed that activity metabolism and digestive costs were competitive within a fixed limit. The limit was thought to be determined by maximum aerobic swimming metabolism, measured by respirometry under precisely defined conditions. It now appears that these components are not strictly additive, as is suggested by the usual formulation in equation 6.3. In fact, this assumption, which underlies much conventional thinking about both organism parts and organisms as mechanical (clocklike!) components of the systems in which they are found, is wrong for some significant combinations of metabolic activities that are of major importance in normal, nonlaboratory environments. For example, when multiple metabolic costs are incurred simultaneously, they act together such that their summary effects can exceed the supposed aerobic limit determined from laboratory tests (Blaikie and Kerr 1996; Krohn 1999). There is no need here to delve into the intricacies of contemporary bioenergetics; the salient point is that the energy costs of securing and processing energy revenues make up a variable but substantial component of the total metabolic expenditure, and that this upper level can vary appreciably, depending on the sets of activities comprehended.

From the point of view of production studies, it is important to appreciate that the metabolic expenditures of organisms are functions of the organization of the system in which they live. That is, changes in metabolism cannot be understood separately from the conditions that conserve the system function of productivity in the environmental context. Particular patterns of activity will influence the ensuing patterns of individual net growth efficiency defined by equation 6.2 and its eventual manifestation at the population level. Such "holistic" concepts were introduced into thinking in ecology at an early stage by G. E. Hutchinson (1953). Among other considerations, in the present context we find it helpful to make use of what he termed the "niche" of a species. According to such thinking, metabolic pattern emerges in the study of ecological *systems* when the organisms are considered as parts of the assemblages within their natural environment, that is, in their ecological niches.

ECOLOGICAL NICHE

We noted earlier in this chapter that Fry (1947) dealt extensively with the effects of environmental factors on the capacities of organisms (primarily heterotherms such as the fishes) to deal with variations in their (primarily abiotic) surroundings. Hutchinson (1957) subsequently generalized this approach to develop the niche concept as an n-dimensional hyperspace depicting the total array of both biotic and abiotic factors that determine the performance of an organism.

Although there is no evidence that Fry or Hutchinson were initially aware of each other's work, it has been a natural development to combine the two approaches, leading to the concept of the "metabolic niche" (Kerr 1980; Kerr and Ryder 1977). In this hybrid construct, the capacity of an organism to function with respect to environmental variation is quantified in terms of metabolic performance capacity within a multidimensional array of external factors. Specifically, the unrestricted capacity of the organism to perform in the absence of competition is defined as the "potential metabolic niche," while the "realized metabolic niche" depicts the organism's metabolic achievements in the real world.

It is worth remarking that the potential metabolic niche of a heterotherm is a variable construct. Within its outermost boundaries, defined as the "incipient lethal limit" by Fry (1947, 1971), the organism, by definition, can persist indefinitely; but its performance potential varies from place to place within these prescribed limits. Characteristically, there is a peak in performance potential for a given attribute (swimming performance or some other correlate of metabolic performance), somewhere near the center of the "zone of tolerance," but this generally decreases near either incipient lethal limit.

This much alone embodies a rich set of ecological possibilities by which to assess the interactions of species, both predators and prey, or their competitors. To choose a specific example, capelin are a common prey of cod. Capelin generally favor colder water than do cod. Accordingly, environmental changes that engender colder water favor the metabolic niche of capelin, at the expense of cod, and the reverse may become true in other circumstances. Clearly, favorable metabolic niche dimensions may conspire to enhance the performance of one species at the expense of another, as conditions permit.

But as Fry (1947) points out, further distinctions are useful. Outside the zone of tolerance just considered, his system of effects of environmental factors identifies a "zone of resistance." By this concept, Fry enables the investigator to encompass the fact that fishes, and other organisms, if they can secure a short-term advantage, can and do enter circumstances that would be lethal to them in the longer term. An obvious example is the lake trout. Lake trout briefly penetrate the thermocline to feed on yellow perch, which occupy the epilimnion of a lake at temperatures higher than the incipient lethal temperature of lake trout. We have reported similar behaviors of Atlantic cod in experimental mesocosms; they briefly penetrate anoxic layers to capture food provided for them in these otherwise life-threatening circumstances (Claireaux et al. 1995).

There is a metabolic equivalent to the zone of resistance that has been particularly well studied for the fishes: the balance between aerobic and anaerobic metabolism. Ordinarily our physiological considerations deal with organisms operating aerobically. However, fishes and other organisms that possess predominantly white muscle tissue (cod being a commercially important example) are capable of accumulating appreciable oxygen debts in the course of meeting environmental exigencies. This capacity has considerable importance both for food-getting and predator avoidance. What is of particular interest here, however, is that this capacity varies among individual stocks of cod. Nelson, Tang, and Boutilier (1994), for example, compared the performance capacities of Scotian Shelf cod with a discrete stock of cod in the nearby Bras d'Or Lakes. Apart from morphological differences long appreci-

ated by the fishermen, they were able to show that the overall performance capacities of the two stocks, in terms of swimming performance
in respirometers, did not differ significantly. However, the Bras d'Or
stock, which may often have opportunity to feed in the relatively anoxic
bottom water layers of these saltwater lakes, exhibited anaerobic capacity to meet metabolic demands that was markedly different from the
nearby Scotian Shelf stock. Anaerobic capacity, accordingly, provides an
additional dimension to the "zone of resistance" identified by Fry
(1947), in the sense that the cod, and presumably other species, use
the capacity to maintain performance in the face of contrary physical
conditions.

There are also recognizable differences in physiological capacity
within stocks. Reidy and associates (1995) have shown that within
stocks some individual cod excel at sustainable aerobic activities, while
others excel at short-term anaerobic pursuits. Just as no human athlete
is gifted at both the marathon and the sprint, neither are individual
fishes: some appear to be superior at one or the other challenge, but not
both. Most individuals, alas, in keeping with their human equivalents,
fall somewhere in the middle ground of not being exceptional in either
regard. Of course, there should be no surprise in any of this; humans are
essentially no less variable. But to the fisheries manager schooled in the
notion of fixed typological species concepts, such differences in capacities of various component groups to adapt to environmental variables
may explain the need to maintain variability in genotypes underlying
physiological capacities.

The details of metabolic niche formulation need not concern us
here; they are set out in the literature, as cited above. What does concern us is recognizing the existence of quantifiable procedures for measuring the capacities of organisms to function in their variable environments. It is this we have in mind when we refer, in subsequent sections,
to the "metabolic niche." We introduce the concept here primarily to
point out that it is a clearly defined, quantifiable performance measure.

The wonder, already expressed in relation to observations of the biomass spectrum, is that with these recognizable differences between and
within species and species components, stable patterns of biomass distribution and organism performance emerge from the measurement of
characteristics of the biological system when these are taken over ap-

propriate scales of observation within different environments. In the case of pattern in the utilization of energy resources by major functional organism groups, for example, we draw attention once more to figure 3.1, illustrating the pattern of overall production efficiency per unit area of habitat found by Humphreys (1979). Such pattern within broad functional groups implies that the total metabolism of the natural assemblage as a component of a system emerges as a nonfragmentable property. In this sense, the apparently flexible responses of individual organisms represent the means of conserving the constant system output through adaptation to the surrounding conditions of their metabolic niches. Thus, the aggregate ecological response is reliably constrained by the limits of energy available per unit area of habitat. It is through this field of flexibility in physiological responses of individuals in accordance with community environmental conditions that realization of the ecological potential production becomes a patterned reality.

POPULATION GROWTH EFFICIENCY

This new, more "holistic" physiological understanding helps to place in perspective the empirical evidence from previous population studies of growth efficiency. It was first shown by Paloheimo and Dickie (1965, 1966b) for fishes that at the population level $K$ exhibits orderly declines as a function of both body size and ration level (which are of course often correlated as organisms grow within their niches). These patterns were termed "$K$-lines" by Paloheimo and Dickie (1965) following the lead of Ivlev (1960). The typical sequence of growth efficiency patterns displayed by the predator population of a given species is shown in figure 6.2.

For a given particle size of prey, growth efficiency of the growing predator declines with increasing ration. The initial period of feeding and growth by larval fishes on small planktonic organisms is, therefore, typically characterized by steep declines in growth efficiency with increasing rations or body sizes. However, growth efficiency for a given predator also increases with increasing food particle size, indicating the importance of "packaging" the energy on the efficiency of predation. Thus, early in their life history, when their swimming abilities have been developed, the small predators shift to an intermediate size of prey, often represented by benthic invertebrates. With further increases in body

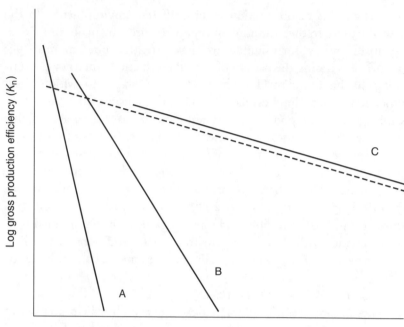

FIGURE 6.2    Idealized *K*-lines for a cohort of fish as it makes feeding transitions through three life history food stanzas, showing the declines in gross production efficiency ($K_n$) at different levels of individual ration ($r$). Line A represents feeding on small particles such as microplankton or, in experiments, hatchery mash. B represents feeding on active particles such as amphipods or copepods; C is for feeding on small fish. The broken line represents the main sequence of efficiencies experienced as the fish grow. (Redrawn from Dickie et al. 1987.)

size, they may also shift again to a diet of small fishes. The sequence of three slopes is typical of what is found under the various feeding regimes that a fish predator sustains during its morphogenetic development.

From the point of view of the overall functioning of the population of an individual species in its niche, however, the general level of efficiency of food utilization can be shown by the dashed line of figure 6.2. This reflects what we have earlier called the "main sequence" of events that lead to the overall position and slope of the growth efficiency lines of the predator as this is played out over its lifetime. It consists of a generally low negative slope with increase in ration, which, for the population in its natural setting, is highly correlated with body size.

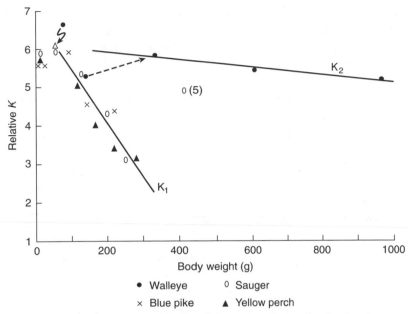

FIGURE 6.3    Relative $K$-lines calculated from growth data for the four large percid species of Lake Erie (about 1920–1950). The outlier for sauger, marked 5, is based on only five specimens and may be in error. The arrow indicates transfer of walleyes from $K_1$ to $K_2$ efficiency lines. (Redrawn from Kerr and Ryder 1977.)

Kerr and Ryder (1977) extended this analysis of growth efficiency to the percid community of Lake Erie. Their observation series (figure 6.3) began with fishes of large enough body size to be taken by the capture gear. The fishes had already made the transition from larval feeding to predation on intermediate sizes of prey; therefore, the data show only two of the typical $K$-lines. Of the four percid species sampled, three of them exhibited a common $K$-line of intermediate slope, while the fourth, a larger-bodied terminal predator, had made the transition from invertebrate to fish prey and exhibited the frequently observed final life-history $K$-line of low slope. Thus, the relationships deduced from feeding experiments on single species were confirmed as applying to an assemblage of species of various body sizes occurring together in their natural setting.

These characteristics of the conversion efficiency have been confirmed for a variety of other situations, as well as for species other than fishes. For example, Conover and Lalli (1974) concluded that the $K$-line relationships between growth efficiency, ration intake, and food particle size for the specialized marine pteropod predator *Clione limacina* helped account for the importance of particle size selection by the predator in maintaining growth efficiency through a major part of the life cycle.

This feeding pattern is in keeping with Jones's (1973, 1978) understanding of the organization of feeding niches within the spectrum. He pointed out how predators may maintain growth within particular stanzas of life history by matching their seasonal growth rate with the growth rate of an abundant or preferred prey. There seems little doubt that such patterns of relationship within the whole community need to be borne in mind if we are to understand the remarkably stable outcomes of the integration of community energy transmission despite local differences in the perplexing details of its structure. For understanding how the increasing productivity results from increasing rations, such observations show that, coincident with increases in ingestion of prey by predators, correlations in predator-prey growth rates play an important role by acting to maintain relatively constant body-size ratios between them.

These patterns of growth efficiency, observed at the population level, establish that the physiological and ecological attributes of predation may both be represented in population models by allometric coefficients. This is the reassurance that we required for the allometric fittings of $K$ in the equations for the spectrum in chapter 4 (equation 4.6). The results of feeding experiments and observations also make clear that these allometric relations will be different for the different scales over which body-size data are aggregated. Relatively high allometric values of the conversion parameter $K$, corresponding with the high slopes of $K$-lines, would apply to the predator-prey relationships maintained between the successive domes in the spectrum, as when fish predators in one dome are feeding on zooplankton prey in the next smaller dome. Within domes, however, it appears that when large fish as predators feed on smaller fish as prey, there is a marked reduction in the slope of the conversion parameters, that is, low allometric coefficients. Hence, within single domes of biomass density there seems to be little change in the efficiency of energy transfer with changing body size of the predator or in the body-size ratio between predator and prey. In this sense, body-size ratios between predators and prey within a biomass density dome are not nearly as critical to the overall efficiency of energy transfer as are relations between domes. From the point of view of this analysis, $K$ may thus be said to truly function as a system level parameter, taking values that appear to reflect the different physiological and ecological levels of the structural features of the biomass density spectrum.

## The Grazing Coefficient, $F$, and Its Interaction with $K$

Ivlev's (1960, 1961) original experiments with the factors affecting growth efficiency take on an additional importance for understanding the predation process. These experiments again explicitly concern fishes. However, they are likely to be applicable more broadly among all organisms that, during various periods of their life history, utilize a change in the range of prey body sizes or types. Their importance in the present context rests primarily in Ivlev's demonstration that a change in the concentration of prey organisms of a given size has an effect on growth efficiency of the predator that is equivalent to the effects of a change in prey body size. That is, the physiological and behavioral mechanisms underlying the growth realized from a prey resource respond to the various properties determining its availability to the predator.

While this conclusion is fascinating in its own right, its importance here is that the concept of an allometric expression of the growth efficiency coefficient, $K$, at the physiological scales of data analysis is difficult to distinguish from behavioral effects that we would ordinarily assign to the predation parameter, $F$. That is, this single set of observations provides us with a basis for allometric expression of a combined food uptake and conversion parameter, $KF$, that we have adopted in the spectral equations. However, the effects of body size on $F$ and their interaction with $K$ go beyond this and can profitably be examined in more detail.

For example, Kerr and Martin (1970) studied the predatory fish (lake trout) of small lakes. Their observations showed that a change in the body size of prey—in this case involving a shift across trophic positions from feeding on zooplankton organisms to feeding on small fish—gave rise to a distinct increase in the growth efficiency of individual predators. Their study showed, moreover, that the increased efficiency of capture by the predator, due to the larger particle size of prey, was sufficient to offset the increased system production costs of a predator that captured the lower numerical densities of the larger prey. It is clear that while $K$ and $F$ may be measured separately in this instance, they have joint effects in their ecological actions on the production system.

In this case, however, Kerr and Martin observed an important additional effect. While the growth efficiency measured for individual predators was increased, the number of larger predators decreased because of trophic considerations, so that the more efficient growth production per individual was mediated through a roughly constant production per unit biomass density. Thus, while the internal details of interaction were changed, the overall production efficiency of the system—hence, possible harvest of fish biomass from the lake—was roughly the same for both lengths of food chain. In other words, conventional wisdom asserts that thermodynamic loss at each successive transfer in a food chain should favor higher production at lower trophic levels. If transfer efficiency increases, however, with improved predation efficiency at lower body-size ratios, the two countervailing tendencies may balance. From the angler's point of view, the important fact was that the fish taken from the longer food chain were of larger individual size. From the point of view of system bioenergetics, this result recalls the constant productivity found by Cyr and Pace in their study of the variable size-composition structures in the zooplankton community (see chapter 3).

We have designated the predation mortality coefficient as $F$, in parallel with the conventional definition of mortality imposed on fish populations by their main predator, the fishery. It is difficult to measure fishing rates in exploited populations and even more difficult to measure predation rate in nature. Their values are not well known for natural populations. Ivlev's (1961) demonstration of the confounding effects of decreases in the average density of prey with prey body size is an important indication of how interactions within the term $KF$ may mean that effects in $K$ are compensated for by a change in $F$, leading to an observed near constancy in the effective slope of the population $K$-line. Few other details are known about it, so the observations of Kerr and Martin (1970) are particularly important for understanding the generality of aquatic ecosystem functioning at the level of the whole system.

We do also know, however, that the rate of population food consumption per unit area decreases with predator body size at about the same rate as the population metabolic rate per unit area (Dickie, Kerr, and Boudreau 1987). It appears, therefore, that within the biomass domes representing trophic positions of the spectrum—that is, at what we have earlier termed the secondary scaling of the spectrum—the combined conversion parameters may result in an allometric slope that is approximately the same as that of the metabolism of the organisms involved.

A somewhat similar conclusion seems to apply at the level of the primary, integral scaling of the spectrum, that is, between the biomass density domes. Dickie, Kerr, and Boudreau (1987) concluded that growth efficiency appears to decrease by about $-0.25$ between the average body sizes of domes representing successive trophic positions—that is, for an average feeding size-ratio, $R$, of about $10^{-4}$ or $10^{-5}$. On this basis it appears that the allometric slope of the combined conversion parameters has about the same value as has been observed for the specific production itself.

From the above conclusion—that there is a constancy of production density functions within each of the different scales of the aquatic biological system—it follows as an interesting corollary that while organisms have many different available strategies for exploiting their ecological situations at different life history stages, the final resulting conversion of energy throughout the food chains is remarkably finely tuned to the capacities of biological materials to act as energy carriers. This is in keeping with the general conclusion of Fenchel (1974), which accords with the earlier results of a study of metabolism by Hemmingsen (1960), that there is a characteristic common value for the efficiency of energy use over the whole range of multicellular organisms in the animal kingdom.

## Conversion Efficiency, Metabolism, and the Body-Size Ratio—$R$

These results return us directly to considering the effects of changes in the values of the composite parameter, $KF$, in relation to values of the parameter $R$, the predator-prey body-size ratio. According to the model of the size spectrum developed in chapters 4 and 5, $R$ is of critical importance in determining various spectral properties. We recall, in particular, that the horizontal positioning and curvature of both overall and secondary biomass domes are functions of $R$ (equation 4.13). Translating that finding to the physiological level of explanation, it is clear that $R$ operates through particular relations to both predator uptake and growth efficiency. That is, the empirical shapes and positions of biomass domes must be partly explicable in terms of efficiencies reflecting the regulation of physiological energy flows that depend on *R*.

To make these relations explicit, we return to the spectral formulation of the relation of production to the ratio of biomass of predator and prey in equation 4.3, or 4.9:

$$P_j/B_j = K_j F_j B_i/B_j, \tag{4.9}$$

where we have recognized that both the production density ratio on the left-hand side and the biomass ratio on the right-hand side have allometric relations to metabolism.

If we consider the case of a constant body-size feeding ratio, we can write the relation $w_i = Rw_j$, and rewrite equation 4.9 in the form

$$\frac{B(w)}{B(Rw)} = \frac{K(w)F(w)}{P(w)}, \tag{6.4}$$

where $P(w) = P_j/B_j$. In logarithmic form, the biomass ratio may be rewritten in body-size terms from equation 4.13 as

$$\log B(w) - \log B(Rw) = \log \alpha + \gamma x, \tag{6.5}$$

whereas from equation 6.4 we now recognize that the allometric expression for the right-hand side consists of two distinct parts: one corresponding to the conversion coefficients, and one corresponding to the metabolic effects of body size on production density. That is, we are now considering that, in equation 6.5,

$$\alpha = \alpha''/\alpha' \quad \text{and} \quad \gamma = \gamma'' + \gamma' \tag{6.6}$$

where, according to the foregoing sections of this chapter, we write

$$K(w)F(w) = \alpha'' w^{\gamma''} \quad \text{and} \quad P(w) = \alpha' w^{\gamma'}.$$

That is, we have a number of coefficients that need to be estimated from the data, and that can provide us with fits to spectra that can then be compared with the empirical data.

While the model fitting is not a trivial exercise, the technology involved is straightforward and is discussed in detail by Thiebaux and Dickie (1993), and by Sprules and Goyke (1994). We need only note here that values for the coefficients of $P(w)$ can be derived from empirical plots of the biomass data compiled by body-size intervals, or from independent metabolic studies. The coefficients $\alpha''$ and $\gamma''$, defining the parameter $KF$, are determinable from equation 6.6 and equation 4.13, which also provide estimates of the three required additional coefficients (log $B_0$, $c_0$, and log $w_0$).

In figure 6.4A and 6.4B we show plots of the body-size dependence of the parameter $KF$ on body size of the predator, given two values for the body-size ratio: log $R = -3.0$ and $-4.0$, estimated from the original, empirical data plots. The corresponding values are listed in table 6.1. Both the figures and the table are taken from Thiebaux and Dickie (1993), who derived them from a fitting of the model to data for the fish from a small lake. The fitted biomass density dome appeared to have a mode at a body size of about 1.5 log kcal equivalents.

We recognize that values of the food conversion efficiency, $K$, must have an upper limit of 1.0, although empirical data, like that illustrated in figures 6.2 and 6.3, have shown that it is rarely higher than about 0.2. Values for $F$ are much less certain, but for theoretical reasons a sustained value greater than 0.5 is highly unlikely, and values of about $0.2-0.4$ have been derived from explorations of empirical data by Dickie (1976). It thus appears that values of $KF$ around the modal body-size length in figures 6.4A and 6.4B are within the expected range. The figures show, however, that at predator body sizes smaller than the modal sizes found in the lake, a predator feeding on other small fish would have to have impossibly high rates of capture or exceedingly high growth efficiencies to maintain itself, because of the low densities of prey that occur under the outer limbs of the spectral dome. That is, the relationship between $KF$ and $R$ shows that efficiency considerations dictate that below a rather narrow range of prey body-size values a given feeding linkage could not be maintained. At these smaller body sizes within the predator size range, a predator operating on a given value of $R$ would find it advantageous to feed on prey likely found at much higher densities in the next smaller biomass density dome.

A comparison of figures 6.4A and 6.4B also indicates the way in which changes in the value of the relative body-size $R$ may be critical to

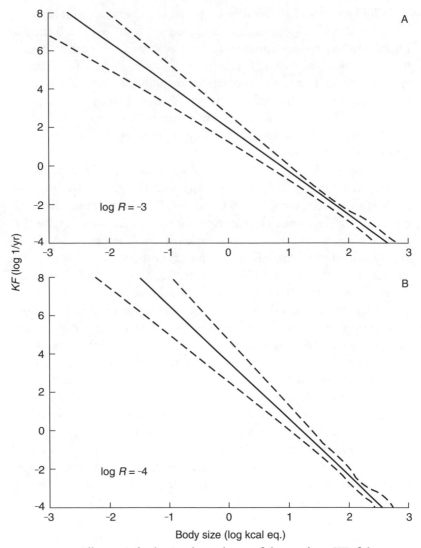

FIGURE 6.4   Allometric body-size dependence of the product *KF* of the predator's combined gross production efficiency and grazing coefficient found for the small lake ecosystem. Standard error limits are indicated by broken lines. Feeding size ratio log *R* fixed at (*A*) = −0.3 and (*B*) = −0.4. (Redrawn from Thiebaux and Dickie 1993.)

TABLE 6.1 Parameters of the fish dome, measured by Thiebaux and Dickie (1993) from fittings of the spectrum of biomass densities in a small lake (Chadwick 1976), and estimates of curvature and position that correspond to two values of the prey-predator body-size ratio $R$.

| Parameter | Value | Units |
|---|---|---|
| $\log B_o$ | $-0.45 \pm 0.18$ | $\log(m^{-2})$ |
| $c_o$ | $0.67 \pm 0.12$ | $\log(m^{-2}) / \log^2(kcal)$ |
| $\log w_o$ | $-0.81 \pm 0.20$ | $\log(kcal)$ |
| $\log a'$ | $0.47 \pm 0.17$ | $\log(yr^{-1}) \log(yr^{-1})$ |
| $\gamma'$ | $-0.29 \pm 0.11$ | $\log(yr^{-1}) / \log(kcal)$ |
| $\log R$ | $-3.0$ | |
| $\log a''$ | $1.85 \pm 0.65$ | $\log(yr^{-1})$ |
| $\gamma''$ | $-2.29 \pm 0.39$ | $\log(yr^{-1}) / \log(kcal)$ |
| $\log R$ | $-4.0$ | |
| $\log a''$ | $3.64 \pm 1.08$ | $\log(yr^{-1})$ |
| $\gamma''$ | $-2.96 \pm 0.51$ | $\log(yr^{-1}) / \log(kcal)$ |

maintaining feeding linkages in food chains. For a predator of given size to maintain its feeding by taking food of a smaller size, such as by an increase in $R$ from log $-3.0$ to $-4.0$, requires a marked increase in the slope of the relationship between the conversion parameter and body size. That is, higher ratios of $R$ significantly decrease the tolerable range of body sizes below which a given density of predator can maintain a reasonable energy yield from the distribution of prey densities expected within the biomass spectral density system.

The set of calculations underlying figure 6.4 also illustrates the fact that as the predator reaches larger body sizes, it is no longer the efficiency of conversion that limits predator densities. That is, if a large predator can find sufficient densities of large-size food, its own population density would be expected to increase in direct proportion. This result, juxtaposed with the apparent importance of the limits of energy efficiency at the smaller relative prey sizes, indicates that there must be a strong negative feedback relationship maintaining the spectral biomass dome structure. As we noted in chapter 5, if a predator on the mode of a spectral dome were to feed on prey that were *near* but not precisely on the mode of the corresponding dome of prey organisms, it would have smaller densities of prey than are required by a simple fixed parameter model to maintain its own density. The continued existence of the spectrum would thus depend on possible variations in the physiological processes that underlie $K$ and $F$. To a certain extent these may be reflected in the many different feeding strategies that have been developed by different species. In addition, however, as we shall see in subsequent discussions, we would expect statistical variation in the value of the predator-prey body-size ratio, $R$, to result in a somewhat smoother spectrum than we would calculate on the basis of fixed parameters.

There are not yet many data that permit us to directly assess the importance of variations in these spectral parameters for maintaining whole systems. In chapter 5 we reviewed in some detail the study by Sprules and Goyke (1994) on the biological systems of Lakes Ontario and Michigan. Their fittings of data to Model II of Thiebaux and Dickie (1993) provide us with an additional estimate of the relative values of the metabolic and conversion coefficients, hence enable us to compare the value of parameters employed in developing the equation system with the value of those found in application elsewhere. Thus, recalling from equation 4.13 that the curvature of the biomass density dome can be written as

$$c_0 = -\gamma/\log R,$$

equations 6.4 and 6.6 permit us to partition the allometric coefficient $\gamma$ into its two parts: $\gamma' + \gamma''$. Sprules and Goyke estimated the metabolic coefficient $\gamma' = -0.31$ for the zooplankton of Lake Ontario, compared to the value of $-0.29$ found by Thiebaux and Dickie (table 6.1) for the fish population of the small lake. There seems to be strong accord between the two situations, both of which are also in agreement with the results of many earlier investigations of metabolic parameters in both laboratory and field situations.

Our present interest, however, rests in the information on the lesser-known coefficient $\gamma''$. The fact that the curvatures of the biomass density domes were quite different in the two situations leads to rather different estimates of this remaining parameter. In Lake Ontario there was a broad distribution of organism body sizes within the zooplankton dome, which gave an estimate of overall parabolic curvature of $c_0 = -0.37$. That is, the distribution formed a distinctive but relatively shallow dome of biomass density with body size. By contrast, the fish dome for the small lake had a relatively small range of body sizes, perhaps reflecting the commonly observed fact that large animals require large volumes of living space. Data fitting in this case resulted in a sharper dome with a parabolic curvature, $c_0 = -0.67$. In addition, log $R$ was estimated to be about $-2$ for the zooplankton of Lake Ontario, but $-3$ for the fish in the small lake. In combination, these estimated differences permit us to calculate, according to the above equation, what appear be relatively large differences in the allometric measure for the conversion coefficients. An estimate of $\gamma'' = -1.05$ obtains for the zooplankton but $-2.29$ for the fish.

With so few estimates of the allometric coefficients from natural data, it is obviously premature to devote much attention to interpreting these two different values. However, it seems worth noting that the high value for the small lake, reflected in the relatively steep slope of the relationship between conversion efficiency and body size shown in figure 6.4, may not be typical of most natural situations. That is, the high slope found by Thiebaux and Dickie appears more likely to have been a reflection of the limited range of predator strategies available in the small lake size they studied, rather than a reflection of a limitation on predator conversion strategies in general. It may be that the values derived by Sprules and Goyke are closer to the normal, suggesting that predator strategies in nature may have a wide range of adaptability, given sufficient environments within which to operate. In chapter 9 we study another situation from the marine environment that suggests something of the full range that may characterize natural systems.

The overall details of precise fittings to real data are an important matter. We have pushed this question as far as we think possible on the basis of the direct data currently available, which we regard as sufficient to support a persuasive argument on the general nature of the relationships. In chapters 9 and 10 we return to the study of comparable parameters that have been derived by Duplisea and Kerr (1995). Conclusions drawn there on the basis of the suggestions offered here confirm that these more detailed studies of the biomass spectrum have far-reaching implications for management and merit further practical study.

## Summary and Conclusions

We can now draw some conclusions about the nature of production systems in relation to physiological variables. Clearly, aquatic production systems do not function in any way like conventional clockworks. On the contrary, aquatic production systems show every indication of adaptive performance within the *KFR* parameter space identified by the Thiebaux-Dickie model. It follows from this premise, at a minimum, that physiological processes figure prominently in the performances of ecological production systems. We conclude from our review that the data justify an allometric formulation for the physiological-behavioral coefficient *KF* that we employ in the predator-prey model of the biomass spectrum.

The physiological differences in the capacities of various fish stocks (and other biota) to respond to the variety of parameters that affect their performances have scarcely been considered in this chapter; a comprehensive examination would require a book of its own. Our intention here has been to provide an introduction to appreciating how the complex, adaptive, array of processes and interactions that support aquatic production do interact in such a way as to ensure a stability of the integrated response. We hope that enough has been said to underscore this important point. We shall have cause to refer to it again, in later chapters.

These considerations imply the concept of the harmonic ecological community. This concept embodies the notion that natural communities that have co-evolved for a long time will have incorporated the adaptations necessary to ensure their ability to function together as integral entities (Ryder et al. 1981; Ryder and Kerr 1990). In this view, niche contention will have been minimized, and major dislocations of local interactions of the production process will have been obviated. The concept of harmonic ecological communities is clearly consonant with the general appearance of orderly biomass spectra in major ecosystems throughout the world. One of our most important motives for continued study of the underlying physiological phenomena is to further understand the degree to which such mechanisms can continue to maintain the resilience of exploited systems to interference by disruptive, external influences. It is the central problem of the final chapters of this book.

༃

# CHAPTER 7

# System Properties of Biomass Spectra

Consistent patterns of behavior detected at a given level of observation provide evidence of system integration: the system is responding to factors acting at corresponding levels of its hierarchical organization. This phenomenon was explored at different levels of aggregation of observations in chapters 2 and 3, and led to the development in chapters 4, 5, and 6 of a theoretical basis for the physiological and ecological scaling of the trophic organization underlying the remarkably orderly biomass density structures that emerge in the body-size spectrum of aquatic ecosystems. That is, aquatic production exhibits general patterns of size-dependent behavior that are clearly consonant with a hierarchical concept of system organization, based on identifiable elements related through familiar energetic processes. Accepting the biomass size spectrum as evidence of a fundamental ordering of internal system behaviors enables us, in this chapter, to begin the necessary final step in a logical analysis: assessing the structure and behavior of aquatic ecosystems in relation to external factors. This final step in considering how aquatic ecosystems reflect the influence of external as well as internal controlling factors is groundwork essential for management of aquatic production systems. In subsequent chapters we address the problem of appropriately applying these considerations to the management of fisheries.

In a formal sense, the Thiebaux-Dickie (T-D) model provides a complete and sufficient explanation of the observed structure of the aquatic biomass spectrum in terms of an initiating trophic uptake and distribution mechanism that is internal to the system. The relative importance of the external context raises a residual set of questions that we have not yet explicitly addressed. For example, while our compilation shows that biomass size spectra are conservative characteristics of aquatic production systems, evidence of somewhat similar discontinuities in the distribution of body sizes has been demonstrated in terrestrial environments as well (Holling 1992). What does this appearance of parallel phenomena in a fundamentally different setting say about the importance of the external environment?

Holling (1992) believed that the late discovery of regular body-size structure in terrestrial environments resulted, at least partly, from the assumptions of investigators that continuous distributions will be found—hence, resulted from the tendency of investigators to aggregate and analyze data at the scales that best reflected their own expectations. The same tendency may have operated in earlier aquatic studies. However, Holling attributes the discontinuities that he has identified in terrestrial ecosystems to generating functions related to external, environmental or "landscape" structures rather than to the internal trophic interactions that we consider causal in aquatic systems. Later in this chapter we will examine in more detail the basis for this difference in explanation. For present purposes, the most relevant point is that while Holling invokes external structural factors, their effects stem primarily from their influence on food-resource availability and distribution. That is, the "global" environmental factors he identifies actually act at the scale of local processes. Thus there is an apparently common appreciation that factors influencing production and distribution of biomass in the community are related to the supply and availability of energy resources. We may therefore turn directly to considering the effects of the organization of external and internal factors in determining ecosystem state, and later return to considering how these may be reflected differently in aquatic and terrestrial communities.

In addition to the fact that discontinuities in biomass distribution appear within both aquatic and terrestrial ecosystems, we know that these structural similarities arise in quite different environments within each type of ecosystem. Similarities that appear despite external ecosystem differences seem to indicate relatively unimportant interaction of the external factors with the internal dynamics. Moreover, these similarities at ecosystem scales are evidently based on quite different species compositions within both aquatic and terrestrial systems. Does the hierarchical nesting of causes in ecosystem organization mean that causes at one level of organization are independent of causes at another? Does a hierarchical nesting of effects mean that internal and external factors will be reflected separately or differently at the different levels in the system? These are questions to be addressed here.

The theory of energy transmission developed by Dickie, Kerr, and Boudreau (1987) and elaborated in Thiebaux and Dickie (1992, 1993) explains how interaction in a complex of biological entities can give rise to stable and orderly patterns in biomass distribution. The unexplained differences at the species level must therefore be related to causes at a different, species-specific level of the hierarchy. So an understanding of the variable species compositions means identifying the level of organization at which production phenomena arise, as well as the nature of the causes. We thus need to understand the relative roles of external and internal factors at various levels in the remarkably ubiquitous general forms that emerge from the complexity that appears when one considers phenomena at the level of observation of individuals in biological communities. We also need to establish whether the patterns we recognize in the hierarchy of organization of the basic elements can be used in applying a knowledge of biomass structure and dynamics to environmental and fisheries management.

## Types of System Behavior

These basic questions of system behavior appear, in fact, to be related to system complexity. In the terms offered by Allen and Starr (1982), apparent complexity and inherent unpredictability of detail are particular properties of middle-number systems. This means that systems observed at a given level of detail may be neither small enough to reliably admit the conventional process of computing output from the dynamics of individual particles (organisms), as is usual in particle physics; nor large enough to permit meaningful conclusions to be reached by taking ensemble averages, as in thermodynamics. Thus, where the main interactions occur at the scale of middle-number systems, neither of the major conventional approaches that have been so successful in physical and chemical analysis in the past can guarantee an understanding of the aspects of ecological system output that concern us.

Rosen, in the first book in this series (Rosen 1991) and its sequel (Rosen 1999), drew attention to this problem of analysis in biology. To phrase the problem in his elegantly explicit terms, conventional physical analysis (e.g., Newtonian mechanics, quantum mechanics) embodies what may be called syntactic precepts; that is, it is concerned with the formalisms underlying the successive states of what are regarded as homogeneous, structureless particles that make up the system—in classical physics, specifically the position and velocity of such particles. The very repetitiveness, or what Rosen calls the recursiveness, of events relating successive states is embodied in the *syntactic* rules of the "laws of physics." It is these rules that form the basis, at this most reduced level of observation, for the stability of explanation of the behaviors of similar physical particles observed at intervals of time in their environment.

By contrast, Rosen points out that biological analysis, necessarily based on the interactions of fewer, often heterogeneous particles, may lack recourse to this essential recursiveness in physics. The behavior of heterogeneous groups of particles is necessarily governed by different kinds of rules, and analysis becomes concerned more with the various functions or relationships among the distinguishable groups of particles—that is, with interaction among the *components* of the system—than with relations among the arrays of the similar particles within groups. By analogy with the kinds of rules governing relationships between parts in a language, these functional or organizational relationships between components within systems are called by Rosen its *semantics*. As he points out, in understanding biological systems we usually have to deal with the outcome of the functions and interactions of the components on the output of the whole system, hence with the semantics of that system. In middle-numbered systems, such as biological systems, it is differences in the organization of the functions and interactions among components that most often determine their characteristic behaviors.

Consider, for example, the active site of an enzyme, a pivotal property of living systems. Its activity depends as much on the tertiary structure of the molecule—the positional or architectural structures that shape the arrangements of electronic charge resulting from the conformation of elements—as on the chemical properties of the component atoms. This organizational pattern determines the potential for energy transfer at a given active site. The potential for energy transfer might be realized by more than one possible molecular architecture, but in specific situations there is generally a limited number of configurations able to support a given effect. The situation is similar in a trophic system. Predator-prey associations in the whole system are mediated by the array of interactions made possible by the range of characteristics of disparate metabolic niches. But it is the patterns within particular niches that, in their totality, determine the potentials for energy transfer. In both cases, the output of interaction in the system entails a semantic potential that is divorced from syntax. That is, the output of the system cannot be explained solely in terms describing the syntactic properties affecting average individual particles.

The conclusion that biological systems may not be intrinsically reducible to single causes via the reductionist program of contemporary physics (Rosen 1972; LeShan and Margenau 1982) has a major impact on the forms of analysis that are appropriate to their manipulation and management. This is not to argue that biology must therefore fall back on vitalism, as was suggested by nineteenth- and early-twentieth-century philosophy. Of course, the biological system necessarily exists as an integral part of the physical universe and obeys its laws. Rosen's arguments instead make the point that customary (syntactic) procedures for physical analysis often are of little importance in understanding what actually results from the organizational (semantic) properties of biological systems. In his terms, many of the really interesting and pressing questions we currently ask about biological systems lie outside the methodology of much contemporary physical analysis, in a kind of terra incognita that physics has yet to fully explore. In Rosen's words, "contemporary physics remains too special to accommodate the class of material systems we call organisms" (Rosen 1991:37).

## A Physical Analogy

To clarify the terminology and significance of these two forms of relationship, we turn to a problem in the physical sciences that is remarkably analogous to these biological middle-number problems, and like them, requires for its resolution a line of analysis that is still under active development. We are referring to a fascinating challenge in system analysis popularly known as the "sandpile" problem, the main characteristics of which are well known to anyone who has played at building sand castles on a beach.

A heap of particles, such as a volume of sand grains, acquires interesting properties that may be regarded as examples of what Prigogine calls self-organizing systems (e.g., Prigogine and Stengers 1984). That is, sandpiles exhibit average properties of slope that are determined by some common properties of the sand, such as texture, grain size, and adhesions due to moisture content, acting in the general gravitational field that affects all the grains. As grains are added to a heap of sand, avalanches occur. The magnitude and frequencies of these avalanches quite remarkably maintain the average slope of the heap, yet individual avalanches are unpredictable on the basis of the common properties of particles alone. In the aggregate, the occurrences of avalanche magnitudes follow an inverse power law, but as every adventurous sand-castle builder knows, an individual event cannot be predicted from the general law.

This condition results from the complex internal (semantic) structures—regions of cohesion in the face of the gravity field, due to moisture for example—that develop as an organization of components in the growing sandpile, creating in the overall or average composition of the sandpile a set of heterogeneous components that varies in time and space. When the values of parameters governing the cohesive forces of a component group at a particular site in the pile fall below critical levels, an avalanche occurs as a particular response of the component group to the general gravitational, syntactic forces. Remarkably, these same variable internal structures are now understood to sometimes lead to the collapse of major containment structures. In some agricultural areas, large structures such as silos and grain elevators have been known to collapse without warning—in retrospect, apparently because of the differential forces exerted upon their walls by sudden internal shifts in the arrangement of the heterogeneous submasses of the particles within them. In the aggregate, heaps of particles of grain are insufficiently numerous to exhibit homogeneous, fluid behavior. However, neither do they behave as solids, much to the discomfiture of the engineer charged with the design of a safe grain elevator.

The complex and labile internal structure of the interactions within the sandpile reminds us of the nature of the stability in organization of the structuring of biological systems, which is why the sandpile problem can be seen as a physical equivalent of the middle-number problem in ecology. In the case of the sand grains, much of the general observed behavior of the heap derives from the internal frictions between grains, augmented by the common behavior (the inertial properties) mediated by existence in the gravitational field. But these have also to act in the presence of variations in moisture and temperature, which give rise to regional differences in the expression of these general forces. In the case of codfish (and other biota), the internal dynamics of trophic interactions are determined by properties embodied in the metabolic niche potentials. These properties are directly indexed in the behavioral and predator-prey interactions that we have already considered in the previous chapter, which together determine the spectrum of biomass with body size in the aquatic ecosystem. At the same time, there may be shifts in the whole biomass species composition, mediated by local differences in the relative size, densities, or behaviors of component predator-prey pairs. These are impossible to predict from the overall, general, metabolic considerations alone. The behavior of the middle-number system thus results from the interaction of forces that act in at least two very different ways: (1) a set of general conditions applying to all particles in a system, but (2) comprehending other special sets of conditions that act to form groups of particles in such a manner that, from time to time or from one position to another, they may appear to be heterogeneous and act distinctively within the average behaviors determined by the set of general forces.

## Critical System States

The sandpile problem thus becomes an intriguing metaphor for a deeper issue—one whose significance goes much beyond metaphor. The seemingly inexhaustible variety that is revealed in detailed study of biological phenomena will have causes operating at one scale; and yet this variability may be comprehended by a stable and predictable functioning, which results in viable organizational structures that emerge only at a different level of observation. Clearly we are observing phenomena that have causes operating at different hierarchical levels, or even in different hierarchies of causes (O'Neill et al. 1986). In each case, however, we need to distinguish the syntactic forces that determine general features of the particles from the internal, special, semantic forces that effectively distinguish heterogeneous components that may act independently within the general system. While there may be no a priori reason why the two types of causes at any given level have to be related, in any attempt to manage or control systems that can be described in this manner, one needs to establish whether there are conditions under which interactions can arise. Bak, Tang, and Wisenfeld (1988) make the general observation that systems of this middle-number type, "having both spatial and temporal degrees of freedom, are commonplace in physics, biology, and the social sciences." They include ecological systems in their lexicon on the grounds that "ecological systems are organized such that the different species 'support' each other in a way which cannot be understood by studying the individual constituents in isolation." In the terms used in our ecosystem analysis, this could be a restatement of the fact that it is the relational aspects of the parameters $K$, $F$, and $R$ in the general processes of energy transmission between species that make the T-D model such an apt, middle-number case in point. As was noted in chapters 3 and 5, it is necessary to be aware of the different scales on which the analysis needs to be applied. There will be one set of values operating at the integral scale, another between groups within the general set. But according to the distinction drawn by Rosen, both levels of aggregation will be subject to the two types of forces, syntactic and semantic, acting on them.

Bak, Tang, and Wisenfeld (1988) also point out that in dynamics theory such apparently complicated systems tend to reduce to a few degrees of freedom if the system's dynamics is appropriately phrased. Therefore, as a means of dealing with the dynamics of the resulting middle-number systems, they introduce the concept of "self-organized criticality" (SOC). In their terms, self-organized criticality may be *the* "underlying concept for temporal and spatial scaling in dissipative non-equilibrium systems." This follows from the fact that these systems evolve naturally toward a critical state, without respect to the specific time or length scales at which the system exists; sandpiles will act as sandpiles whether they are large or small. Furthermore, over a wide range of size the whole system supports component groups that will cause the manifestation of "strange" behaviors.

It follows that an essential feature of an ecosystem characterized by SOC dynamics is that its manifestations at any level of observation will be governed by local (semantic) driving forces acting on, or between, heterogeneous components within the general (syntactic) field created between individual biological elements and their environment. By and large, the idea of local driving forces appears to accurately reflect the mechanics of ecological system functioning. For example, as we have already pointed out in chapters 2 and 3, while the level of energy transmission in a trophic system may be associated with the general density of predators and their prey, it is the local configuration of the prey-predator interactions that determines the availability of particular body sizes of prey to a predator. Hence, while overall production is reflected in average biomass densities, the actual rate of transfer of energy between component groups in the system is determined by local factors such as the distributions of biomass with body size. This may abruptly change with variations in density or size distribution of either predator or prey components without significantly disrupting the whole system. However, there will be limits that are not always easy to predict and that reflect the tendency of the output of the system to exhibit self-organizing dynamics.

As an important specific example, we may consider mortality rates for a fished population of a community. Overall mortality can be measured as a global average over the entire habitat of the stocks and used to explain its relative abundance in relation to rates of fishing. Nevertheless, the rate of uptake of the fish as prey for the fishery—hence the actual mortality and the ensuing production in the system—depends on the parameters of local scale distributions or "schools" of both the prey and the fishing vessels as predators, and will vary locally with variations in densities within the population-wide distribution. The result is often quite different from the expectations of average uptakes based on assumed uniform distributions (Gulland 1955; Paloheimo and Dickie 1964). This means that, while the field of dynamics that describes a stock may use the overall mortality rate as a syntactic descriptor, controlling this average rate does not necessarily affect the mechanisms that actually control the local energy transfer dynamics. Controlling the average mortality may not affect production in any particular stock or species in any predictable way. In fact, one must expect in such systems that, even with no change in the identified overall parameters or of the external constraints that act on them, because of the self-organizing dynamics there will be relationships among some components that will always give rise to unexpected system states—what Holling (1986) called "surprises." We deal with some observed practical consequences of this problem in chapter 8 in the interpretation of fishing mortality and fishing effort distributions.

In the same vein, there have been many arguments and analyses in population studies, especially in fisheries, that have attempted to relate the abundance of brood or spawning stocks to the success of new year-classes of progeny—the so-called stock-recruit problem. The variability of both the stock and recruitment data for species has often been a source of frustration to researchers and fisheries managers alike, who are convinced that the logic of an overall relation must be manifested in some way but that the evidence for it is buried in details of the scatter in the data of system interaction. This view appears to be supported in general by the plethora of factors, both local and general, that can be seen to act on survival at different times in the life histories of different species, or on the same species in different environments. In this case it becomes clearer, from the inability to predict the resultant recruitment, that the average or global measurement of a population parameter will often fail to provide an adequate index of the rates of interaction in the local processes that actually act on the heterogeneous system components and that determine the outcomes of the natural processes.

Revelations about the different scales on which interactions in ecological systems take place, and debates about whether they are related to internal dynamics or external environment, are very well known. Yet, because of the simple, single-species models that are used in fisheries, the different scales and types of interaction as causal aspects of system behavior have been consistently neglected in the drive to explain them. This failure has become most obvious in recent years in how the actual results of fishing have deviated from the output of models constructed to predict the effects of recruitment and mortality coefficients over the distribution of "stocks." The results often tempt investigators to conclude that environmental factors must be to blame. What is new in our consideration of the multispecies, hierarchical models implied by the biomass spectrum is an understanding of system behavior that views fisheries ecosystems as middle-number systems. From this perspective the factors operating at the various levels may act syntactically as constraints on interactions between average elements (as may be the case with the level of nutrient supplies), or as semantic factors affecting the organization of energy transfers among the components of the system (as in predator-prey dynamics). This perception of alternative system behaviors at quite different scales leads to quite different perceptions of the causes of change and the appropriate management controls.

Thus, where the patterning of body-size elements in a biomass spectrum emerges as a stable structural feature of an exploited community, it becomes reasonable to suspect that interactions governing its behavior, what we call its causes, cannot be ascribed to global or syntactic factors alone. The further realization that management measures must operate at the same scales as the factors they are designed to modify gives this line of argument particular force in problems of management. In the language of system theory, any hierarchical system will reach critical states that cannot be defined or explained only in terms of the global or syntactic averages acting at one level in the system.

In general terms, the foregoing observations about the properties of middle-number systems in ecology mean that what have been termed self-organizing properties must often prevail in biological systems governed by local, semantic factors. The behavior of such systems may appear unpredictable, even chaotic, in the face of changes in state that alter the parameters of the field of cohesion of the stocks, although the global parameters describing the overall community dynamics seem to change little; or even in cases where the observed general level of exploitation seemed to be subjected to "effective" control or manipulation. In the same way that grain elevators built to hold huge masses of cereals may collapse from shifts in internal forces among heterogeneous components, generalized global management procedures applied to ecosystems may give the appearance of being ineffective—indeed, may *be* ineffective—if, in their action on particular stocks, they do not identify and affect the causal dynamics at the appropriate scale.

This is an important proposition, not to be overlooked in times of obvious failures in managing exploited natural ecosystems. For that reason, we devoted considerable attention in previous chapters to examining the nature and scale of ecological interactions. Given the basically local nature of the *KFR* parameter space specified in the T-D model for any level of interaction, all the important, organizational interactions in this biomass spectrum model of natural ecosystems may need to be seen as short range and local in their effects. This neglected point of view embodies substantial consequences for the provident management of aquatic production systems. In the language introduced by Rosen (1991), the semantic structure of ecological systems may be the most important to their function; their syntactic relations are likely to describe a structural framework that constrains the whole system production, but production processes themselves entrain a series of local feedback processes, which in most cases need to be specifically and separately constrained to ensure the maintenance of a balance within the system.

The effects of these two types of causes are readily comprehended by analogy with the characteristics of sandpiles. They are more difficult to see in the details of the dynamics of biological populations. For this reason we turn our attention more specifically to a review of various examples of biological systems. We begin with a simple computer model of a simulated fisheries ecosystem that meets the criteria of a middle-number system that we have used above. The results verify the kinds of interactions expected. We follow this with a review of actual details of structures found in examples of various natural systems—some simple, some rather more complex. All reflect the kinds of variations that the foregoing review of the theory of natural system interactions leads us to expect. Finally we return to considering the general properties of natural systems as they are reflected in biomass spectra.

## A Simple Computer Model of an Ecological System

Wilson and associates (1991) constructed a unique, multispecies fish community simulation model that illustrates some of the consequences of interactions in middle-number systems where there are both general and local internal driving forces acting among the heterogeneous components. The results of this simulation allow us to examine aspects of the different effects of global and local properties and the ways that they appear in the fisheries context. The simulation was constructed in the form of an age-structured multispecies system. Five designated fish species were assigned different conventional parameters of growth, mortality, and reproduction. However, the model additionally assigned what was termed a "community predation" function (Sissenwine 1986), in which the adults of each species were enabled to feed on the young-of-the-year of any species. It is by this predation feature that the authors designated their model as an interactive multispecies model, rather than an aggregated single-species model.

Two versions of the model were developed. In one of them the equilibrium abundance of individual species was allowed to become established according to the "thermodynamic" balance of individual species population parameters. That is, in this version there was an upper limit to an individual species population abundance based on its general dynamic parameters, but effectively no interactive community predation on the young of individual species. In the other case, an overall food limitation was imposed on the system so that before the equilibrium population abundance of any given species was reached, predation by the total adult population on the array of recruits to any species reduced the abundance of young so that the resulting total population abundance did not exceed the limit of the imposed "carrying capacity." That is, in the second case, individual species abundances were limited by predation and mortality imposed by the entire community on the sum of the newly spawned year-classes of all the species. The two models were envisioned as tools for studying the role of community predation on species composition in a system such as that of Georges Bank, where a variable abundance of species is supported in a relatively large, but apparently energy-limited, ecosystem.

The different results of the two systems were examined in relation to four features that its designers believed to be invariable characteristics of natural fish production systems, and which we have already come to expect in middle-number ecological systems: stable total biomass; unpredictable biomass of individual species; energy efficiency indicated by species compensation within the stable aggregate biomass; and lack of an obvious relationship between spawning stock and subsequent recruitment within individual species.

While no simple simulation model can be expected to capture all of the complex behaviors that natural systems are capable of exhibiting (Fogarty 1995), the authors nevertheless found that these particular four behaviors emerged in the second interactive case—even with quite different values assigned to the global characteristics of the various species, or to the overall level of the total energy supply to the system, as long as the limited carrying capacity induced interaction among the species. That is, the four features showed up as the apparent combined consequence of the syntactic condition of a limited food-energy supply, combined with a system output that depended on a mechanism of energy transfers governed by internal, local, or semantic trophic interactions among the populations of the species. The specified properties did not appear when this interactive community predation was absent. Where the individual species were allowed to reach an equilibrium dictated by their individual average population parameters, individual species remained constant at relative levels dictated by their species dynamics.

In the view of Wilson and associates (1991), the observed time sequences of behaviors of the individual species in their interactive model production system provided evidence of what the researchers termed deterministic chaotic behavior. In this model, while the overall biomass was relatively stable, the representations of the individual, component species were variable and apparently unpredictable. The assumption that the results can actually be technically described as chaotic was questioned by Fogarty (1995). However, whether the system output was merely variable beyond the possibility of realistic prediction, or whether it exhibited the characteristic mathematical features of chaos, is mostly irrelevant to judging the significance of the model to our present study. The essential point is that simulation of an energy-limited system linked through local community predation processes gave results that displayed predictable overall output, based on variable, unpredictable internal details of interaction among the component species. It thus produced behavior reminiscent of observed fish production systems, and displayed properties resembling the behavior of middle-number system dynamics. In effect, results of the model by Wilson and associates provide a simplified ecosystem realization of interactions of the two kinds of dynamics that appear so clearly in our experience of the sandpile problem: interaction of components defined and organized at the semantic level, constrained within a general syntactic framework.

It seems reasonable to conclude that in this example we are dealing with another case of the characteristic conservative patterns of behavior by systems whose syntactic dynamic properties are underlain and actually controlled by a pattern of events that acts on the various components in a well-defined local parameter space. The result in this limited instance of simplified species interaction appears to correspond to characteristics of behavior that have been repeatedly observed in exploited fish assemblages. We therefore need to seek further evidence that real systems of this type—in particular, ecosystems in which the body sizes of the individuals are also distinguished—will have the coherence of interaction that gives rise to biomass domes in a real system, whether underlain by simple or complex species interactions. For that reason, we now reexamine the evidence for the conditions under which specific structural patterns in the size-dependent organization of fish subdomes arise in the overall spectrum. We need to be reassured that complex but stable system generalities follow from the measurable local system interaction properties that are defined in the T-D model as the parameters $K$, $F$, $R$ and in the body-size domes corresponding to fishes, and that, as with sandpiles, these generalities possess properties that emerge without relation to the particular time and space scales of the whole system in which they occur.

## Characteristic Natural Systems

We are making potentially strong assertions about aquatic ecosystem behavior. If they are true, orderly biomass spectra should appear as ubiquitous properties of trophically related natural aquatic systems, operating in accord with the parameter spaces defined on $KFR$ and as described in chapter 5. That is, they result from actions that are at scales different from the generalized syntactic scales of the properties of organisms that have heretofore occupied the almost total attention of fisheries system management models. The substantial body of published evidence of ecosystem density structure in chapters 2 and 3 revealed formal evidence for characteristic biomass size spectra. To examine the scale of origin of these structures requires observations taken with greater resolution than applies to the global properties under which ecosystem behavior has been portrayed up to now. We introduce here evidence from both complex systems and from some of the simplest aquatic production systems we know of, or can imagine, in which the nature of these interactions is exhibited more clearly. In the simpler situations, drawn primarily from freshwater communities, we can focus initially on the emergence of biomass subdomes in fish communities comprising one, or at most a very few, terminal predators at various levels of varying global dynamics properties. These can then be compared with the properties displayed by the more common, but more complex, ecosystem examples that dominate marine production systems, of which the Wilson and associates (1991) model is a heuristic simulation.

To be considered successful, a theory must make new predictions that are testable. The theory we have outlined to this point admits various testable predictions. Most obviously, it predicts that fish biomass domes should be the general pattern to be observed. In particular, because the theory is fundamentally rooted in the semantics of the *KFR* parameter space at any level in the system, it predicts that biomass domes should occur independently of the spatial scale and species diversity of the production system. Because of self-organizing dynamics, it also follows (within broad limits) that biomass domes should occur independently of the exploitation intensity the system experiences. In the following sections, we test those predictions by comparative study of examples of differing complexity.

## *Unexploited or Lightly Exploited Single Stocks in Freshwater*

Johnson (1972, 1994 and numerous references cited therein) has specialized in reporting the structures of unexploited populations of various northern lakes of low general productivity. He refers to them as "reference systems" because of their manifest simplicity (and for thermodynamic considerations not pursued here). These depauperate oligotrophic systems show domes of biomass density in relation to body size that are in close accord with the fish domes we have described in earlier chapters, but often based on a single predatory species. Johnson documents numerous instances of such domes in northern lakes fish production systems (as well as in other special types, such as tropical bivalve populations). His repeated observations support our conclusion, as described in chapter 5, that biomass domes will appear wherever a particular *KFR* parameter space prevails. These domes are similar in general shape and position to those found in the multispecies population of a small inland lake by Chadwick (1976), described by Boudreau, Dickie, and Kerr (1991) and discussed in relation to the Great Lakes study of Sprules and Goyke (1994) in chapter 5.

Fish growth patterns are broadly subject to genetic strictures, but their realizations in any situation are determined additionally by environmental factors expected to affect both predator and prey sizes. In general, as in figure 7.1, Johnson's studies show unimodal fish body-size distributions throughout the northern lake systems, mostly based on the

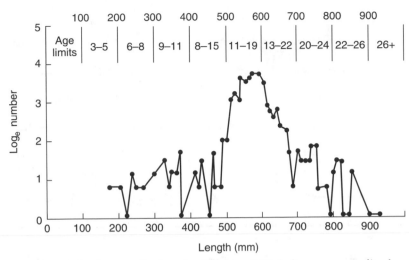

FIGURE 7.1 Catch curve for length of lake trout (*Salvelinus namaycush*) taken in the summer of 1962 in Keller Lake. Note the weak correlation between body length and age. (Redrawn from Johnson 1972.)

same few fish species throughout. However, in some cases, apparently where there is some additional "stress" in the form of fishery removals, there is an increase in the frequency of replacement recruitment, which results in the establishment of a separate, pre-recruit dome of smaller fishes of the same species. In other instances, for the predatory fishes only, as in figure 7.2, a secondary dome also becomes evident to the right of the primary modal dome—reflecting a shift of the predator to larger foods, thus instituting a supplementary *K*-line of lower slope, as was described earlier for more southerly lakes by Kerr and Martin (1970, and see below). In some circumstances, this supplementary dome is at least partially supported by cannibalism, as is the case with the Ogac Lake cod stock discussed below. However, in general, there is no clear evidence of the importance of cannibalism in comparable "reference systems." It may be more common than has been recorded (Campbell 1979; Sparholt 1985).

These studies indicate that neither the presence nor general shape of domes of biomass density with body size is a simple function of the general level of fertility of real aquatic systems, nor are the domes an exclusive property of biological communities that are rich in fish species. The

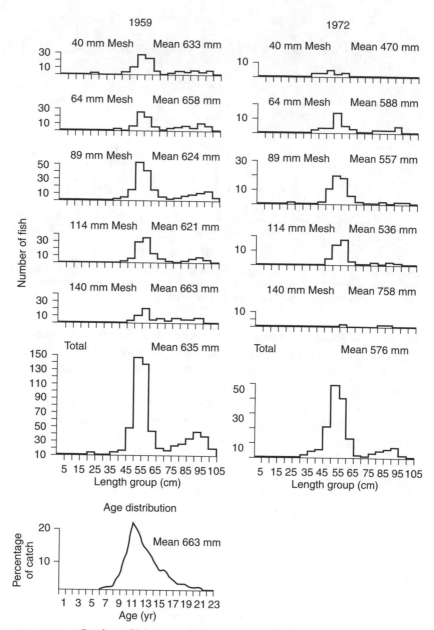

FIGURE 7.2   Catches of lake trout in 1952 and 1972, Lac La Matre, Northwest Territories. Essentially the same bimodal size configuration is obtained in each year. (Redrawn from Johnson 1994.)

supplementary domes that show up in some situations should likely be regarded in the same light as the secondary domes within the general fish body-size distributions that were demonstrated by Sprules and Goyke (1994) in the southern, more complex, and, in some instances, relatively more eutrophic Laurentian Great Lakes. That is, the domes at larger body size seem to depend on the community adapting to a sufficient body-size gap between two groups, one of which can then serve as food for the other. In the case of the dome at smaller body size, this recruitment must depend on an improved food supply of smaller organisms, as well as an improved egg production that results from reductions in the densities of adults due to modest exploitation in these generally impoverished lakes.

## Exploited Single Stocks in Freshwater

Martin (1966, 1970) has provided valuable, comparable observations of lake trout (*Salvelinus namaycush*) production systems that occur in a large set of oligotrophic lakes in central Canada. These lakes are all subject to angling fisheries, of varying degrees of intensity. Some of the contained stocks are planktivores, because zooplankton is essentially the only suitable prey available in their lakes. Others can avail themselves of larger prey (forage fishes) as growth permits. These piscivore production systems differ substantially in that the growth patterns and ensuing size compositions of the trout-planktivore populations exhibit much smaller body-mass-at-age than do the trout piscivores. Overall productivity (as indexed by annual fishery yield per unit area), however, does not vary appreciably among the lakes, despite the differences in food chain length (Kerr and Martin 1970).

To rephrase these findings in terms of the T-D model, the relatively smaller prey size available in planktivore lakes causes the trout growth efficiencies to decline rapidly with increased predator size (i.e., the steep initial *K*-lines), holding the fish biomass dome (composed entirely, in many instances, of the single species of lake trout) at a small average size relative to the biomass dome exhibited by piscivorous populations of the same species elsewhere. In lakes where larger prey are available (prey fishes), or in instances where cannibalism occurs, the larger lake trout are able to shift to the secondary, lower-sloped *K*-line afforded by the larger prey. The more sustained growth efficiency leads to larger absolute body sizes of the predators in the community (Kerr 1979); hence domes of biomass density range over larger body sizes. However, productivity itself, indexed as fishery yield, remains roughly independent of these effects, as it did in the Wilson and associates (1991) model. Apparently the higher turnover at the smaller body sizes completely compensates for the smaller absolute biomass at these sizes. That is, only the size composition of the realized production is affected by the addition of a biomass dome at a new position, not the total productivity of that system.

To grasp the full significance of these observations, we need to return to the discussion in chapter 6 of the importance of physiological properties of organisms. That is, it needs to be appreciated that the fishes, like many aquatic poikilotherms, possess remarkable plasticity in responding to variations of biotic and other factors that affect their production capacities. Martin (1966) demonstrated strong association between prey size and the body sizes attained by predatory lake trout (figure 7.3), but perhaps the most convincing evidence that the ecological consequences of growth plasticity of lake trout are mediated by the availability of larger prey types (forage fishes) was provided by Martin's (1966) experimental transfer of planktivore lake trout to a lake where the piscivore opportunity existed. The transplanted planktivore individuals uniformly responded with accelerated growth that rapidly closed on the growth and body-size patterns of the resident piscivore population.

Observations of the dependence of fish growth patterns upon the sizes of available prey have been made in a number of other studies. Kriksunov and Shatunovsky (1979) report such relationships for the smelt, *Osmerus eperlanus,* as do Evans and Loftus (1987) for *O. mordax.* Reports are particularly prevalent among the freshwater salmonids, where observations are perhaps most obvious because of the general simplicity of the species assemblages (Campbell 1979). Sparholt (1985), for example, documents several instances in Arctic char populations (*Salvelinus alpinus*). However, the literature is replete with comparable examples of prey-predator effects in nonsalmonid species (e.g., Holčík 1977; Bagenal 1977). Again, we are seeing examples from nature that resemble the situation modeled by Wilson and associates (1991) in the simulation for Georges Bank. The overall productivity of these systems is constrained by global properties, but the dynamic processes through which this production is mediated are under the control of local inter-

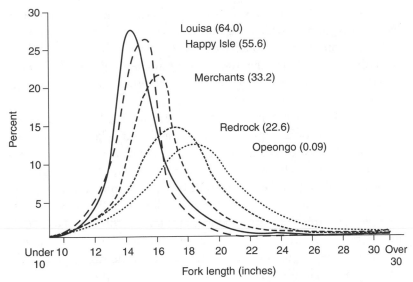

FIGURE 7.3    Length composition of lake trout in five Algonquin Park fisheries for the period 1936–1964. Percentage of lake trout feeding on plankton shown in brackets. (Redrawn from Martin 1966.)

action events, among which the most important appear to be the trophic parameters underlying the predator-prey relationships. Where the body-size information is available, these local interactions can be seen to give rise to domes of biomass density among both the predators and their prey. These cases show effects comparable to the species effects studied by Wilson and associates (1991), but we now have information on the community responses by body sizes as well. In the aggregate, the results support the predictions of models of the body-size spectrum of the biomass.

## A Simple Unexploited Marine System

Marine fish communities do not commonly offer the same economy of simplicity that is afforded by Johnson's "reference systems"; hence evidence for the development of fish biomass domes in marine communities is more difficult to unravel than it is in freshwater systems. Marine communities possess generally longer histories, and often offer a greater diversity of resource species on which a community complex can be established. There is, however, at least one marine fish community that does qualify, and in which the biomass patterns correspond with the freshwater communities noted above. We describe it here, albeit briefly, to put to rest speculation that there may be inherent differences between freshwater and marine fish species or communities that would be relevant to our present considerations.

The stock we have in mind is the population of Atlantic cod (*Gadus morhua*) that occupies Ogac Lake, a virtually landlocked, marine fjord on the south coast of Baffin Island. The cod stock itself has been described by Patriquin (1966), and its habitat by McLaren (1966, 1969a,b). At the time of their studies, it was unfished, apart from occasional visits by the Inuit; to our knowledge, this may still be the situation. In brief, the Ogac Lake stock of young cod is supported initially by predation on invertebrates. *Gammarus sp.* are the most important, and are apparently available to the youngest cohorts. Subsequently, the major part of the growing cod stock subsists on a diet of seemingly inappropriate prey (for cod), consisting particularly of sea urchins. As might be expected from subsisting on food of such low caloric density, these cod fare poorly, and their slow growth patterns reflect the low energy density of their prey resource. In consequence, the principal biomass dome of the Ogac Lake cod is small—centered at about the one kilogram body-size level (figure 7.4). This modal size is at least an order of magnitude less than cod stocks of comparable age exhibit in other environments, even when moderately fished. A notably weak association between age and size is evident in the population.

There is, however, a lesser component of the Ogac cod stock that is cannibalistic on the major part of the stock. The biomass dome of this second component is shifted much to the right, ranging across a larger span of body sizes (figure 7.4). Predation on smaller fish evidently affords cod the same advantages enjoyed by Martin's piscivorous lake trout. We conclude that Atlantic cod, when undisturbed by major fisheries exploitation, are quite capable of the same measures of trophic adaptation as any freshwater species of fish. Accordingly, it seems safe to conclude that marine and freshwater species of fish do not differ in any important regard for our considerations.

The significant difference between marine and freshwater ecosystems appears to result from their historical provenance, which, in the former case, entails a much longer opportunity for historical assembly,

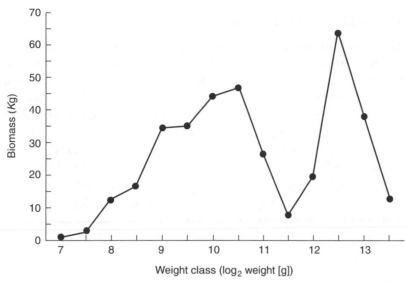

FIGURE 7.4 Bimodal biomass size spectrum for the cod of Ogac Lake, Baffin Island. The large cannibals to the right are supported almost entirely by predation on the smaller cod. (Plotted from field notes from 1952 and 1957 provided by I. A. McLaren. For associated references, see text.)

and doubtless co-evolution, of the species assemblages. A similar conclusion was reached by Warwick (1984) as a result of his comparisons of benthic communities from different environments, and parallel observations have been made by Duplisea (1998) at various points in and near the Baltic Sea. In any event, the ensuing result is much the same; the integral size composition of the species-rich assemblage of a marine fish community may look much the same as that of a depauperate freshwater system, in the sense that biomass domes appear to be the rule. The resultant form is not unlike what arises in a Fourier analysis of the power distributions of ecological phenomena observed over a wide band of frequencies (Platt and Denman 1975). In these analyses, the aggregate form of the distribution of biomass among numerous individual species domes is reflected in an overall general structure of the biomass spectrum at body sizes that reflect different metabolic turnover rates. However, the same structures appear in various production systems, despite differences in scale and internal complexity, apparently dependent on the same critically important local energy transfer mechanisms.

*Exploited Large Fisheries Systems*

Comparable evidence for biomass domes in large, complex fisheries can be drawn from several sources. The biomass structures for a single year in two of the Laurentian Great Lakes, showing multiple domes from plankton through to fish, have already been illustrated in chapter 5 (figures 5.2 and 5.3). This is for one year only, but the data are for an ecosystem range of organism body sizes. Subdomes for the fish component of the Scotian Shelf ecosystem compiled by Duplisea, Kerr, and Dickie (1997) over multiple years are shown in figure 7.5. This series depicts the remarkable uniformity of the fish dome, which is known to have consisted of a variety of different species in different abundances over the period of observation. Similar data for the fish subdomes have been recompiled by Duplisea and Kerr for the North Sea and Georges

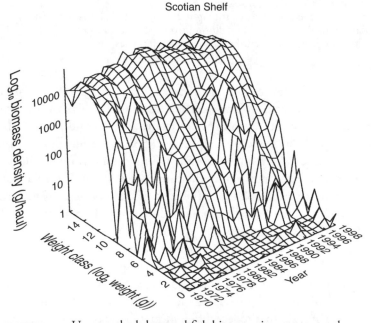

Scotian Shelf

FIGURE 7.5   Unsmoothed demersal fish biomass size spectra on the Scotian Shelf of Atlantic Canada from 1970 to 1998. (Data were replotted, with additions, from the smoothed figure in Duplisea et al. 1997.)

North Sea

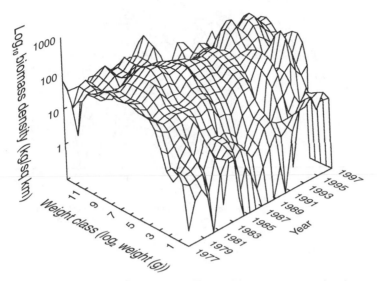

FIGURE 7.6 Unsmoothed demersal fish biomass size spectra for the North Sea, 1977–1997. (Data were derived from the English trawl survey and kindly provided by Dr. Daniel Duplisea, CEFAS, Lowestoft, U.K.)

Bank for the purposes of this book, and the results are shown in figures 7.6 and 7.7. We note that the biomass concentration values shown in these figures are internally consistent, but are not directly comparable between figures.

Regardless of differences in the survey protocols for these various systems, including the areas and volumes swept by the survey vessel gears, the designs of the surveys, and the expressions of concentration for the resulting catches, there appear to be strong similarities in catch per standard haul or estimated concentrations of fish among all three of the communities. That is, differences in survey techniques notwithstanding, the shapes of the resulting plots are evidence for the existence of remarkable regularity in the body size compositions of the fish communities in these complex, heavily exploited fisheries production systems.

Georges Bank

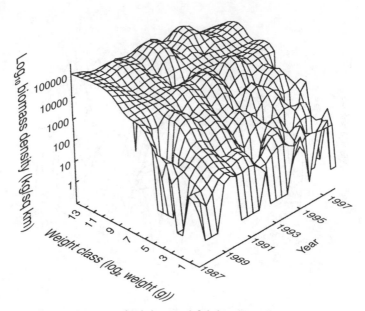

FIGURE 7.7    Unsmoothed demersal fish biomass size spectra for Georges Bank, 1987–1997. (Derived from the Canadian groundfish survey.)

Quite apart from the similarities between the major sea fisheries and smaller, less diverse fisheries, the regularities exhibited by these three major sea fisheries are interesting in themselves. In particular, all three exhibit the characteristic domed structure expected on the basis of the Thiebaux-Dickie model. However, it is noteworthy that the three systems appear to exhibit somewhat different aspects of the internal processes of species replacement expected to lead to the common, domed property that is observed. For example, the Scotian Shelf system exhibits somewhat more prominent subdomes than the other two. In this ecosystem, despite marked variations in abundance of individual species within the relatively stable overall abundance, Duplisea, Kerr, and Dickie (1997) found no evidence of major patterns in the replacement of one species group by another over more than two decades of observation. They concluded that the term "diffuse replacement" best described the observations.

In contrast, changes in the North Sea and on Georges Bank have been marked by consistent changes of the abundance of major groups, among the pelagics (herring and mackerel, for instance), or among the gadoids (cod and haddock), and even among the elasmobranchs. On Georges Bank, for example, the larger and valuable gadoids were largely replaced by unmarketable elasmobranchs such as spiny dogfish and skates, maintaining the overall form of the fish dome but to the detriment of the fishery.

The situation in the North Sea is somewhat different, and might even be considered worrisome. In recent years the largest size groups of fishes there have conspicuously diminished in abundance, and while abundances of the smallest size groups have increased, possibly suggesting new recruitment, it may be wishful thinking to suppose that these comprise the requisite species to maintain the fishery. The alternative view is that the traditional North Sea fishery is at the brink of commercial collapse.

Clearly, biomass domes are ubiquitous within the major, complex, multispecies fisheries, both marine and freshwater. Interestingly enough, moreover, the data suggest that the domes occurring in different areas cluster around the same ranges of body sizes, evidently even while based on different species. As with the Wilson and associates (1991) model, the individual species structure is quite variable between systems, and within systems from time to time. However, the stable biomass density domes are maintained over relatively long series of years, despite changes in both the rate of exploitation and the species compositions.

Taken together, the complete set of examples provides ample verification of the predictions we set out earlier in the chapter. Fish biomass domes are indeed a common property wherever observations have been made, and, in accord with theory, their existence is apparently independent of the spatial scale of the production system, its species richness, whether or not it is exploited, or indeed, whether it is freshwater or marine. These important general results provide us with the material needed to begin a reassessment of the production dynamics made possible by this additional new knowledge. Spectral characteristics of these important fisheries systems and their relations provide the material necessary to formulate a more effective management. Duplisea and Kerr (1995) developed a specific methodology that may allow us to utilize these observations of variations in spectral structure. We review their results in relation to management objectives in chapter 9.

## General Properties of the Biomass Density Spectrum

We return to the question of the interaction of what we earlier called the semantic and syntactic factors acting on different levels in ecosystems, as they influence the patterns in the biomass spectrum that persist despite very marked differences in environments. These differences may now even be extended from purely aquatic to terrestrial environments, through the review by Holling (1992). We have already concluded that the trophic transmission mechanism, postulated in the T-D model, appears to offer a robust explanation for the main features of all of these patterns: namely, that they seem to depend on the semantic scale of predator-prey feeding interactions. So it is hardly surprising that their effects appear at the appropriate scales of observation in so many different communities of organisms despite the diversity in the general environmental circumstances. Why, however, should the terrestrial studies have reached a basically different conclusion about the mechanisms responsible?

The most obvious explanation lies in the general fact that ecological phenomena are unlikely ever to be the result of only one cause; hence the trophic dynamics explanation must be capable of refinement that admits the influence of other factors. At the outset, for example, it should be recognized that, as described in chapter 4, the T-D model is, at base, the description of a type of feeding that is particularly common in aquatic systems: particulate feeding. Selective particulate feeding behavior is the one that most obviously yields the gaps in the body-size distribution of biomass on which the particular, oscillatory nature (across body size) of the model solution depends. While particulate feeding has a remarkable ubiquity in aquatic environments, the creation of jumps in energy flow between predator and prey applies a fortiori in situations where the particles captured are so small that they require a filtering, rather than an individual capture, mechanism. From this point of view, the ubiquity of the spectrum in aquatic environments may be seen as a particular result of the small body sizes of the planktonic organisms or of bacteria that form the base of the food chains in aquatic systems, and that dictate at the semantic scale of observation an orderly sequence of body-size gaps in the energy transmission.

Because of the imaginative experimental and intellectual work of Ivlev (1960, 1961) and Rashevsky (1959), it has also become possible to generalize the particulate feeding model and recognize that part of its wide application must be related to the fact that the energetic efficiency of feeding, which is so directly affected by changes in particle size of the ingested particles, may be affected in equivalent fashion by changes in the degree of aggregation of particles that do not, themselves, change in individual size. Environmental patchiness is a semantic property of natural biological systems that is increasingly recognized as a feature of both marine and terrestrial environments. There thus seems little doubt that the general distributional attributes that control availability account for much of the wide applicability that has appeared for the T-D model of energy transmission in aquatic food chains. It is reasonable to conclude that, in this sense, those aspects of the environment that have an effect on the availability of food to a predator are of first-order importance in determining the occurrence, shape, and position of domes in the observed biomass density spectrum.

In a sense, it is perhaps not surprising that the analyses of Holling, of mechanisms that seem to account for body-size gaps in the biomass distribution of mammals and birds in terrestrial environments, should have failed to identify the significance that we have attributed to the trophic energy flow mechanism. His deductions are based on comparisons of empirical data in which the overall frequency of occurrence, rather than the density, was the principal criterion for setting up the body-size distributions of various mammal or bird species. Throughout the animal kingdom, we would expect body size and density to be related to many factors, in which the very occurrence of suitable food will influence distributions taken over long enough time or space scales. In the case of fishes, for example, the stocks of large cod and similar species have clearly adapted their style of living to vast current patterns that occur over the coastal shelves. In these species it is essential that spawning take place in areas where the currents will carry the relatively passive, pelagic young into favorable rearing locations. These rarely seem to be close to the principal adult feeding grounds. Therefore, the whole syntactic landscape pattern of the availability of resources for successive stages in the life history of the species, as well as the relations between body size, migratory behavior, and energy storage capacity of the body, and the current patterns that prevail within the species distribution, all have to be considered in determining the occurrence and distribution of species.

While this explains how terrestrial body-size gaps may appear to be related to landscape-scale patterns, it hardly obviates the likelihood that the complex life-history patterns of mammals or birds, particularly the large terrestrial herbivores with the remarkable gaps in energy flows from grasses to large body sizes, will also demonstrate the effects of the availability of energy for transmission across these body-size gaps on their densities. Indeed, we have already referred in chapter 3 to the perception by Damuth (1981) of how body size is related to home range overlap in large grazing mammals. That is, his observations at the semantic scale of organism distributions in relation to food requirements yields evidence of an allometric influence of terrestrial organism density on feeding distributions. However, it also seems highly likely that the well-known irregularities of the body-size steps in the energy transmission from herbivores to successive predators in terrestrial food chains obscure the regularity of density patterns with body size that appears so strongly in the aquatic body-size spectra. The investigations of Enquist and associates (1999) into the allometric regularities that show up both in plants and animals in relation to their internal energy transport mechanisms and internal energy dispositions are strong testimony to the physiological rules that generally govern organism growth and survival. It would be surprising if additional ecological effects of feeding densities did not also appear rather broadly. Differences between communities of mammals and birds may cast further light on the relative importance of distributional factors in establishing community patterns of energy flow in relation to organism metabolism and density, in terrestrial as well as aquatic environments.

Even in aquatic environments, various investigators have questioned the significance of the predator-prey models in favor of the continuous flow model of Platt and Denman (1977) through successive body-size categories. Continuous flow models may have a special status among the very smallest of the living particles in aquatic systems where the energy transmitted by absorption through cell walls, rather than particulate feeding, is involved. On the other hand, the relative abundance of bacteria has been reported to constitute a body-size group that is more distinctly separated from the distribution of the larger organisms than, say, the phytoplankton are separated from the zooplankton, both in body-size range and in relative density. That is, the distinction of bacteria into a body-size dome separate from organisms that feed on bacteria may be based on the energy-gap mechanism. The difference in density level may be dependent on the special population efficiency mechanisms that result from the substantial fractions of bacterial populations that are observed to be metabolically inactive at various times.

The evidence is insufficient to reject a hypothesis of the underlying importance of the energy gap in the trophic flow between body sizes on the subsequent distribution of biomass and production. However, it is equally clear that the body sizes of the primary producers in aquatic systems must impose a strong set of initial conditions from which the remaining properties follow. This is clearly an area where the basic nature of the external environment can be seen to impose a rigid set of defining circumstances. We thus agree with the urging of Steele (1991) that future study of the possible bridging of the gaps in the different spatial and temporal scales involved in terrestrial and aquatic systems in particular, through attention to the unifying functional relationships that system analyses represent, has a very high priority if we are to understand the basis for effective protection and management of our biosphere.

Apart from the important influence of body-size class on the production and biomass density patterns of the host production system, there are other obvious questions that appear referable to the external forces acting on the system. For example, in the aquatic system, why should a "fish" dome occupy a different location than a "zooplankton" dome, or an "infaunal" dome, for that matter? The obvious answer is that different constraints prevail, as a function of body size, across the available spectrum of ecological opportunities. For example, even in the early life-history stages of fishes, water viscosity does not significantly determine their survival strategies. Some zooplankton, on the other hand, may well exhibit the strictures of life at a low Reynolds number (Conover, 1978; Mann and Lazier 1991). In terms of the influence of external conditions, this means that aquatic organisms that operate at appreciably different body sizes necessarily experience quite different limitations on their abilities to function. For this and similar reasons, different size classes of organisms exhibit various responses to their external medium. Given these different constraints, which are called into play as a function of organism body size, it is small wonder that these effects translate into different biomass domes, which are apparently based upon predation strategies that depend on adaptation to external conditions.

## Some Conclusions

It may now be more obvious that, in the terms we are using, the commonly measured global attributes of ecological systems, such as their average mortality and recruitment rates, are to be regarded as virtually extraneous, syntactic factors in the scale of their impact. That is, while they may characterize the extensive features of a particular system, in middle-number ecosystems they do not sufficiently define the actual mechanisms that control the system's observed output, although they may interfere with its expression. In managing our natural systems, we neglect such findings at our peril.

In general, it also appears that many other system parameters, especially those that are often considered external to the main dynamics, exhibit both syntactic and semantic aspects in their impacts on ecosystems. Factors such as primary productivity and thermal field set the global parameter space for the production system. But it is also well known that the way these factors act is not specifically identified in the fact of a correlation between, for example, ambient temperature and metabolic rates of organisms. Nor, indeed, are their effects fully explained by organismic physiological processes alone, as is clear from the biomass density effects on production processes underlying the spectrum. Thus, it needs to be recognized that in aquatic systems such as fisheries, the overall level of different environmental impacts—including the mortality caused by intense fisheries exploitation and the growth rates influenced by food and temperature, which are ordinarily treated as causative global factors in production—also needs to be considered at different levels of integration in the systems (Walters and Collie 1987). By analogy with our deductions about the scale of events determining trophic energy transfer, it may be that, given an apparent thermodynamic balance among syntactic factors, the important practical actions of all of these factors have to be appreciated at the local level where semantic influences seem most clearly exhibited. That is, as causes of production change, many "external" system factors may affect local factors that are as significant to the detailed organization and output of the system as they are to the general level of production in the community.

It is especially pertinent in cases such as mortality on stocks to note the deductions of Sissenwine (1984), who estimated that on Georges Bank the fish community consumes roughly 60–90% of its own production, and that most of the consumed individuals are in the postlarval phase of life. Similar dynamics have been implied in measurement of turnover rates in the fisheries of the North Sea and Scotian Shelf (Gislason and Helgason 1985; Dickie, Kerr, and Schwinghamer 1987), and in the data for major fisheries ecosystems reviewed by Bax (1991, 1998). It is obvious, then, that even in intensely exploited populations, where substantial fractions of the production are removed from the system by factors that are clearly external to it, the main pathways of energy transfer are still as much a reflection of internal forces and local scales of organization as are those at unexploited, lower stages of food chains, where interactions are more obviously local and internal to the system. Thus one can reasonably conclude that internal, short-range interactions that are semantic in nature may dominate the organization of energy transfers both within ecological systems and between the natural system and its exploiters. If so, one can similarly conclude that self-organizing processes may come to dominate ecosystem dynamics at certain scales and intensities of interaction.

In general, in the recurrent structures of the biomass spectrum we will find features of the biomass distribution that appear similar from the point of view of the energy dynamics, but may be instances of truly heterogeneous groupings of the component populations, when one takes into account their interactions with the properties of their physical environment. A mechanism—analogous to that in the water viscosity arguments—related to the grain size of the benthic substrate was postulated by Schwinghamer (1981b) to help account for the complexity of dome structures that appeared invariable in different benthic environments. Similarly, it is highly uncertain that the trophic body-size gap is a sufficient explanation for the complicated spatial as well as temporal variations that seem likely to show up in data related to the secondary biomass domes. The reality and generality of secondary domes was underlined by the careful and detailed studies of Sprules and associates (1994, 1995) in the Great Lakes. However, these apparently stable structures required careful statistical identification in the "noise" generated by various influences on the data. If additional considerations are needed to explain recurrent features of biomass domes, we have evidence of the real subtlety of action through which various external influences in ecological systems may be reflected in ecological structures.

For these and similar reasons, we might suppose that similar body sizes can entail quite different predation strategies. Aquatic biomass domes appear to be centered on body-size optima determined by physical considerations. Although many details remain to be resolved, there seems to be no mystery, or even surprise, in any of this. Schmidt-Nielsen (1984) reviews a variety of factors that introduce functional discontinuities as a function of body size. It follows that aquatic biomass domes are a predictable consequence of some simple properties of aquatic ecosystems. The fact that these are not materially altered in shape and position by external factors is not, in its turn, an indication that external factors have no effect on production.

## Summary

In this chapter, we undertook the essential step of examining how aquatic ecosystems reflect the properties of middle-number ecological systems. In particular, we examined the self-organizational characteristics that emerge in a semantic system embedded in a syntactic context— initially in terms of the simple analogy of the sandpile problem, and subsequently in terms of a heuristic system model of fisheries production. In the process we sought to establish some linkages, perhaps speculative, with terrestrial production systems. Finally, we considered some predictions that emerge from such considerations, and concluded there is ubiquitous evidence in fish production systems for the existence of consistent, comparable domes of fish biomass in a wide variety of aquatic production systems.

∾

# CHAPTER 8

# Alternative Views of the System Underlying Fisheries

Fisheries systems are, of course, liable to many influences in addition to those we have been considering up to now. Ideally, commercial fisheries are controlled by economic factors that must establish a positive difference between revenues based on sales of landed fish and the costs of building, fitting, and operating the vessels, equipment, and companies that undertake their disposal. In fact, however, the relations between landings and the motivations for making them may be impossibly complicated by the different economic and social strategies of various countries or the accounting practices of fishing companies. In many cases there has been no immediately discernible relation between the biological yields of fish and the "effort" expended on those yields. That places in jeopardy theories that purport to use feedback relations between profits and costs as potential market controls of future fishing mortalities. We need to assure ourselves that in our analyses of fisheries ecosystems we are dealing with real factors of interaction having a measurable and predictable effect on future production.

This book is not the place to discuss economic control of fisheries systems in detail. However, we would be disregarding our responsibility if we did not point out that failures in management of existing fisheries owe as much to inappropriate or oversimplified definitions of what constitutes the fisheries production system, as they do to errors in the estimation of values for parameters that are held to measure the relations between natural production and its exploitation.

Given the generality and constancy of the biomass spectrum in so many aquatic ecosystems, it is evident that natural production is constrained within a framework of organism types that tends toward full utilization of the basic energy resources available for biological production. In this situation, current fisheries theory entertains a most dangerous oversimplification: that system output can be measured now and projected into the future as the sum of future biological productivities of the principal species present. The single-species models that are applied to measure the relations between mortality, yield, and effort bear little resemblance to the reality of the complex, flexible, hierarchical systems revealed in the spectral data. Single-species models are simply incapable of anticipating or measuring those apparently important interactions responsible for the spectrum and that point to relationships between species, either as competitors for food or in the many predator-prey trophic relations that make up the natural production base.

Our thesis throughout this review is that management objectives in fisheries can be reached because, given the opportunity, fish production systems unfailingly organize themselves into the characteristic biomass and production density patterns revealed in the spectrum. In later chapters we show how sensitively these patterns can be measured using the many survey data that are already available from well-developed and established sampling surveys, such as those summarized in chapter 7.

It must be clear from our previous discussion that the data supplied by these surveys reflect the different types of effects acting at the different system levels. The syntactic levels of influence related to thermal field, nutrient supply, and the like are well known, even if not very precisely determined. What are not well known, or even recognized, are the actions of the semantic influences, which act on the organization of the biological system and include the effects of both natural ecological, trophic factors and the disruptive fishing factors. While there are still arguments about the precise nature of these semantic level interactions, comparing observed changes in fishery systems with the nature of the sandpile dynamics helps to clarify what must be involved. Formulating fishery production effects according to spectral theory thus constitutes a more realistic and complete description of system changes than is available through present methodologies because it accords with the complex nature of the dynamic controls that is implied by the very existence of the spectrum.

In this chapter, by recourse to an understanding of the different types of causes that operate within the biological system or between it and the economic system, we can begin to see why there has been so little connection in recent years between the expectations raised by fisheries conservation regulations and the actual alarming downward trends in landings in so many important fisheries areas of the world. Because of the very nature of systems, as this term is now understood, theory that cannot specifically relate control parameters to the underlying system mechanisms will be unable to extrapolate to the unexpected, new system states, which are always outside the range of recent experience. Furthermore, only in the presence of a hierarchical arrangement of the causative factors in the fisheries system can one recognize the well-known fact in physical system analysis that large-scale phenomena do not constrain the actions of causes that operate at smaller time and space scales. For both of these reasons, one cannot anticipate effects of disturbances in real, complex systems by simplifications based on the presumption of homogeneity of the units in production systems or of syntactic uniformity in the distributions of the biomass densities on which the systems operate.

In this chapter we concern ourselves with three major oversimplifications of current fisheries models. The first two concern the illusory concepts of homogeneity of the properties of the basic production units. The third concerns the consequences of assuming uniformity in the distribution of these units. The three problem areas are

1. the "unit" stock,
2. the related question of the place of species in the production system, and
3. the mode of action of fishing mortality.

In each of these cases, the basic assumption made in existing fisheries theory is that the properties of interaction can be ascribed to the level of syntactic influences. We believe that it is essential to take seriously the evidence that there are important influences of a different type, acting primarily on the organization of the fish community. That is, fisheries are, first and foremost, subject to the actions of what we have termed semantic factors in a hierarchical system of relationships. Recognizing this nature of effects in fisheries system dynamics suggests how interpretation must change to permit us to see the place of semantic properties in ecological dynamics and how they must be treated for effective management.

## The Unit Stock Concept

A pivotal concept of conventional fishing theory, which prevails even now in many of its applications, is that of the "unit stock." According to Ricker (1975), a "stock" is merely the "part of a fish population which is under consideration from the point of view of actual or potential utilization." The conventional definition of a "unit stock" is therefore loosely determined by context, and in subsequent usage has become quite ambiguous in its application. In general, Booke (1981) noted that a unit stock might be described either according to genetic criteria, or phenotypic characteristics. The expectation (or hope?) in applying this definition is that the population of a species within a statistical data-collection district has uniform dynamic production characteristics (e.g., that all cod of a given age group are identical—as alike as so many electrons—and that density distributions of the whole stock do not vary significantly within the defined statistical sampling districts assumed to be occupied by it).

These are curious notions that are unlikely to be either true or useful in most situations. Fundamentally they seem to derive from the typological species concept, still extant as late as the 1940s, that a species is a discrete entity, constant in space and time, and with severely limited scope for possible variation. Sharply contrasting with that concept is the view now commonplace in most branches of biology that a species is a group of interbreeding natural populations that is genetically isolated from other such groups. From this latter view it follows that much internal species variation is possible and to be expected, even to the point of subspecies formation (i.e., polytypic as opposed to monotypic species). Although Mayr (1997) notes that, "Typological thinking is never enlightening in the study of life," it appears to persist in contemporary fishing theory.

While we can accept the utility of the concept of a unit stock in some special situations, such as small freshwater bodies, or among some semelparous stocks where homing to the natal stream is relatively strict and critical to the survival of the stock, we view it as a dubious assumption in most major fisheries. Looked at objectively, the only justification we can find for the continued use of the unit stock concept is as a statistical abstraction that permits the application of simple calculations of average properties to real populations. This kind of approximation might once have been needed when the sampling data were very sparse. It could hardly be justified today if one inquires about the variances of such averages and their meanings.

A different view of the organization of production systems, with consequences that are significant to a hierarchical view of their nature, was offered by Levins (1968). He introduced the concept of "metapopulations," which, in his original usage, referred to an assemblage of local population units, perhaps each attached to some different spawning locale or time, which, taken together, ensures the survival of the metapopulation as a whole. That is, the metapopulation exists without reference to specific physical boundaries, such as the shorelines that define the limits of lacustrine systems, and may exist as individual spawning groups that support different aspects of the total phenotypic or genotypic plasticity, any one of which must have some finite probability of extinction. The importance of the concept lies in the fact that even limited genetic exchange among the units within the metapopulation enhances its overall probability of survival. The implication is that a more heuristic alternative to the abstraction of a uniform unit stock is a balanced interaction of differing subgroups. It is the plasticity and exchange of variety among subgroups, not uniformity, that is important to the maintenance of a given level of productivity.

As we might expect, the initial concept of metapopulation has since been refined and expanded in various ecological contexts (e.g., Hastings and Harrison 1994; Harrison and Hastings 1996). In fisheries applications, McQuinn (1997) has considered the question in detail, with particular reference to the Atlantic herring stocks of the Gulf of St. Lawrence. He concluded that the metapopulation concept of Levins, associated with moderate genetic connection among the constituent components of the herring metapopulation, best accounts for the structures of the Atlantic herring stocks he examined.

Similar conclusions, but with apparently lesser rates of gene flow, appear to account for the stock structures of northwest Atlantic cod. We base this view on the two lines of evidence: phenotypic and genotypic. In the first line, it is evident from the physiological work reviewed in chapter 6 that cod stocks are not all alike. In terms of their phenotypic, physiological capacities, performances within or between samples of the cod stocks are not uniform, some individuals being manifestly superior at aerobic activities, for example, while others are better at short-term anaerobic performance. As we also pointed out in chapter 6, aerobic performance is associated with long-term sustained activity. This is the swimming performance expected of stocks that undergo lengthy seasonal migrations, hence a characteristic of obvious significance in offshore environments where extensive migrations are an important feature. By contrast, anaerobic excellence is associated with the short-term requirements of securing prey or eluding predators, and can be an advantage in other identifiable local situations. Excellence in both categories is apparently not possible, either within or between stocks, or at least is not sufficiently common to be detectable within the laboratory populations drawn from different "discrete" stocks.

This heterogeneity in performance potential suggests that codfish, contrary to assumptions of contemporary management models, are certainly not all alike in their abilities to take advantage of particular environmental opportunities. That is, important physiological trade-offs exist between apparently narrow alternatives that cod and other species must "choose" in order to adapt to challenges imposed by local habitat requirements. Adaptations within subunits of stocks are significant to local survival and production and are characteristics that would have to be maintained in the more extensive and fluid genetic exchanges required for metapopulation survival in more general environmental conditions. That is, stocks are not static entities but result from a balance that is maintained by migration, selection, and genetic exchange within the fluid phenotypic groups of organisms that make up a natural population.

Our second line of evidence is associated with the deeper implications of these phenotypic differences in performance as they become more clearly associated with the genetic evidence. We earlier noted the prominent genetic feedback loop in figure 6.1 to draw attention to the rich potential, among such fecund organisms as many of the fishes, for intense genetic selection that permits adaptation to local conditions. In the marine environment, this local adaptation is at a scale and intensity virtually unimaginable in terrestrial vertebrates. Pogson, Mesa, and Boutilier (1995), using nuclear DNA restriction fragment length (RFLP) loci, were the first to reveal the existence of extensive genetic differences among Atlantic cod stocks, a view that has since been abundantly reinforced, primarily using nuclear DNA microsatellite loci (e.g., Bentzen et al. 1996; Ruzzante et al. 1996; Ruzzante, Taggart, and Cook 1996).

The weight of genetic evidence for cod suggests a close fit to the meta-population framework for stock structure of Atlantic cod, but with an apparently lower rate of genetic exchange among substocks than is deduced by McQuinn (1997) for Atlantic herring. We have focused on cod here because that is the best conjunction of both stock structure and genetic exchange that is currently available. And although citation of absolute rates can be misleading, it is useful to note that Ruzzante, Taggart, and Cook (1996) estimate that the number of Atlantic cod migrants among some of the substocks they analyzed would be roughly 20–30 per generation, low rates that, while at odds with the conventional view of discrete structure of unit stocks (particularly when applied across broadly defined statistical districts), could ensure general survival through genetic exchange between neighboring areas, in balance with the selection pressures required to maintain the local adaptations. That is, the apparently constant production efficiency of a meta-population depends on a remarkable balance between selection and gene flow among the component stock units.

There is thus the important problem of the effects of changes in abundance on survival of subpopulation units. According to what has been termed the Allee effect (Allee 1931), single populations may have a higher probability of becoming extinct when they are rare, because scarce reproductive opportunities may cause negative population growth rates below a critical density threshold (Amarasekare 1998). This phenomenon appears to be known in other contexts as "depensation" effects (Myers et al. 1995), which constitute a statistical recognition of the negative population growth that has been measured at low densities, without specific reference to mechanism (Neave 1954). It follows from such considerations that extinction events as a function of low population density may be more likely at the local scale of the metapopulation, even if general syntactic estimates of density give the appearance of being adequate owing to higher averages calculated from including densities measured at other localities in the population area. Heterogeneity of type necessarily interacts with heterogeneity of distribution in these fundamental questions of biological dynamics. Once again, the distinction between syntactic and semantic approaches to analysis becomes a matter of considerable importance.

There is, of course, the additional important question of how useful characteristics might become reestablished if they should be locally extinguished. McQuinn, in his treatment of Atlantic herring, stresses distinguishing the relative importance of imprinting and learned behavior in assessing the potential for such reestablishment. Classical imprinting to natal situations is the prevalent supposition for fishes, based primarily on salmonid models, but it is likely not the common strategy. We say this because it is unlikely to be flexible enough to be successful, except among the semelparous anadromous Pacific salmons where it has been well studied and is obviously a requisite strategy. In iteroparous circumstances, more effective strategies can easily be imagined. Learned behavior, in particular, offers the short-term flexibility needed to cope with secular changes in environmental conditions.

It is worth pointing out in this context, that while behavioral interactions linked to genetic exchanges between stock components of a metapopulation are clearly important to survival within the species, similar phenotypic plasticity may be a requirement of continued trophic interactions between species. In this sense, the metapopulation concept of survival within species has a counterpart in the important interactions between species that constitute either competitive or predator-prey relations. Their nature and role in production has been equally neglected in single-species models. In offering this observation, we are mindful of the need to consider these two disparate aspects of the organization of fishery production systems together. The genetic component we regard as informational. The trophic component embodies energy transfer, which we have been regarding at the physiological and ecological levels of organization. Despite their different natures, both informational and energetic means of exchange are inextricably linked and operate in close conjunction, particularly in relation to what we have been calling the semantic influences on system organization.

Clearly, the metapopulation concept carries important implications for management of the distribution of fishing effort. If fishing is to be a sustainable economic activity, and if the metapopulation model, or anything similar to it, is the actual basis for continued existence of production in large-scale marine fisheries, then it follows immediately that effort must be regulated by considering the spatial scale of the metapopulation distribution, and that fishing intensity must be sensitive to the needed potential rates of genetic transmission between substocks of the metapopulation. These rates will of course vary with the metapopulations under consideration. Taken together, such considerations can only reinforce our earlier questions about the relative operational importance of semantic, organizational levels of influence, as contrasted with calculations of changes in averages, which are usually taken at the syntactic level as though population events were determined by changes in the numbers of homogeneous, average units. Where in current fishery models the latter basis is accepted as the one through which fishery exploitation works, the significance of the actual fluidity indicated by the evidence of variation will not be perceived. By contrast, on the basis of both energetic and genetic evidence we can begin to question conclusions about future production based on a wrong level of perception of the internal dynamics.

## The Role of Species

Consideration of the role of the metapopulation in survival within species has its obvious counterpart in questions about the significance of particular species in maintaining the production of a community of organisms. The main difference is that survival within the species depends primarily on information transmission, whereas community production requires the action of energetic processes. In fact, differences in the perception of production, according to the numerical methods of single-species population models and biomass spectrum models, derive from different perceptions of the importance of informational and energy models of system behavior. Economic or social criteria introduce additional management requirements, and characterize the variability of the system in terms of economic and social concerns. We have not yet found a way to accommodate an understanding of interactions of these various system concepts in a single, unitary description—a fact that shows up in the problem of predicting species changes.

One of the principal initial reasons for study of the spectrum of biomass with body size was the emergent regularity in its structure and production when the compilations were carried out over species. The advantages of the original compilation by Sheldon, Sutcliffe, and Prakash (1972) were utilized by Schwinghamer (1981b), and later by Warwick (1984), as a welcome solution to the difficulties of calculating production for an environment from the myriad of species encountered in benthic sampling. Holling compiled the data for his analysis on somewhat different criteria than those used for the aquatic studies, but his analysis has now extended this recognition of regularity in body-size domes but variability in species to the mammals and birds in terrestrial communities. Questions of management based on economic or social criteria also require that we characterize this variability in terms common to the artificial and natural production system processes. The data for fisheries-related ecosystems provide for a more detailed examination over relatively long periods. We use the data to frame a clearer perception of the significance of species composition in relation to this required dual formulation of the system dynamics.

Regier (1973) was the first to point to the long-term sequence of various sizes and species that appeared in the course of growing exploitation in a major fishery, on the basis of data for the time series of landings from the Laurentian Great Lakes. The large-bodied sturgeon, lake herrings, and lake trout that supported the early fisheries were gradually replaced by small-bodied species, such as the yellow perch, smelts, and alewives, earlier regarded as "trash" species by the fishing industry. Notably, however, the total annual landings did not change much during these shifts in the species composition of the catches, despite significant growth in fishing intensities. Subsequently, Sutcliffe, Drinkwater, and Muir (1977) showed a similar data series for the more open marine area of Georges Bank, where the relatively constant total production overlaid variability in the relative abundance of an even wider variety of constituent species, despite growth of fishing intensities.

Details of these changes became clearer from presentation of the 25-year research vessel data set of standardized surveys of species density and body size on Georges Bank, with additional information on the North Sea and the Gulf of Thailand by Hennemuth (1979). The data reviewed by Holden (1978) and Daan (1980) also showed the transition in North Sea fisheries from an initial dependence on larger "table" fish to the appearance and growing importance, with increased rates of fishing, of what are called "industrial" fish. Clearly, such observations are ubiquitous, suggesting that, at a sufficiently large scale, the emergent stability of production that shows so clearly in the biomass density compilations is also reflected in the commercial exploitation results.

Pope and associates (1988) suggested that not only did the changes in species composition take place in a seemingly orderly manner, but that some of the shifts, both in the North Sea and Georges Bank, represented species or species-group replacements. In the most detailed data, that for Georges Bank, there even appeared to be two quasi-reversals in the replacement trends. Initially, the most common bottom-feeding species, such as the cod and haddock, were replaced by an increase in smaller-bodied pelagic species, such as mackerel and herring. Later, these were replaced in their turn by a resurgence of larger benthic species. This second time, however, the benthic species composition has been very different and composed largely of commercially undesirable elasmobranchs: dogfish and skates.

Some of the generalizing of species changes into trends of species groups may be a subjective function of the kinds and degrees of aggregation used in the compilations. The biological taxonomic approach, in common with the economic named-species approach, has a degree of flexibility in identifying the species groupings used to investigate relations to aspects of the ecosystem. For example, earlier speculations often argued that the changes were correlated to the physical environment, or to food resource changes. Until Regier's study the importance of the intensity of exploitation was less clear. Partly this reflected the unanticipated stability of total yield, partly a failure to appreciate the scale of density effects that we discuss in the next section of this chapter. Where investigations have been confined to syntactic concepts of system relationship, there has been a general failure to relate cause and effect to the mechanisms involved. Evidence of long-term trends was required before it became accepted that the overall effects of fishing did not accord with the expectations of the sum of individual species responses to exploitation.

By contrast, fishermen, whose concern with profit and loss directs their attention to local changes at the semantic level of population distribution and interaction, have been quick to assign the cause of changes to overfishing. On their local scales of observation, what may appear in statistical calculations as gradual changes in species composition, appear as devastatingly abrupt, and lacking in historical precedent—facts that relate the changes to the uniquely increased fishing pressures. From neither point of view, however, is there any question of the reality of changes in species over periods of years to decades, despite generally more stable total fishery production over the same periods. The problem of finding a common understanding of cause arises because of the differences in levels of information available from the different sources, as well as differences in the scales of criteria considered relevant by the different observers.

A somewhat intermediate view of the degree of regularity of species changes has been offered by Duplisea and Kerr (1995) and Duplisea, Kerr, and Dickie (1997) as a result of their study of data for the Scotian Shelf in relation to the spectrum. We discuss the results in detail in chapter 9, but note here that species variability appeared to be less ordered than on Georges Bank, with no discernible persistent trends of replacement of one group by another, and few persistent correlations among the species group representations. However, two distinct periods of different levels of fishing intensity were identifiable within the total 22-year data set. During the earlier 11-year period of generally higher fishing intensities and lower densities, landings were composed principally of species that have a typically smaller range of body sizes. This was succeeded by a period of lower fishing intensities that followed implementation of the Law of the Sea and resulted in generally higher densities, especially of the larger-bodied, gadoid species. Duplisea, Kerr, and Dickie (1997) described the changes in species composition as a "diffuse" replacement; that is, they did not question the general impression of compensatory replacement, but in their data there was no evident basis for particular species substitutions. It is important, however, to note the apparent linkage of the trends of changing body size and changing species to the rate of exploitation.

The observations of Sutcliffe, Drinkwater, and Muir (1977) and Hennemuth (1979) for Georges Bank were the basis for the model of Wilson and associates (1991) described in chapter 7, designed to examine the hypothesis that changes in species composition might result from semantic-type biological interactions in an energy-limited environment. Their results not only faithfully reflected the experience of constant biomass with apparently unpredictable replacements of species, but also suggested that the ancillary problems of unpredictable replacement and unpredictable trends in recruitment of species in relation to the abundance of mature adults are aspects of the same types of system behavior. Their analyses demonstrated the unique and major importance of interaction of species mediated through local trophic transfers among population components. The dynamics of the model were such that the syntactic limitation on the energy supply was mediated by semantic trophic interactions among the heterogeneous component groups.

Regarding these observed and theoretical behaviors of species as component groups in the behavior of middle-number systems offers a more general explanation for the perplexing species problem in ecosystem rescarch and management. That is, the unpredictable behavior of the abundance of organisms taken at the levels of component species groups—with regularities at the more reduced level of particles identified by the assignment of a body-size criterion in the biomass spectrum—implies a reflection in a middle-number system of the different levels of cause that may affect output. In such systems, the constituent particles play different roles, depending on the level of aggregation at which they are defined. For example, when the biological system is seen to be organized as heterogeneous components consisting of species, the results of interaction are as unpredictable as the avalanches in the sandpile. In the physical illustration, the cohesive forces governed by moisture content are as indefinable in their area or site and time of interaction as the mutual relations of the species groups identified in the natural biological system. The detailed results of interaction of such components are unpredictable on the basis of general dynamics information of either physical or biological systems, despite the inevitable overall energetic outcome.

On the other hand, in the ecosystem, when these same species particles are redefined as body sizes related according to a set of metabolic rules, the results of interaction appear as a regularity in the biomass distribution that is as highly predictable as the slope of the sandpile. In system behavioral terms, the biomass spectrum description of the system, through the use of a well-known set of metabolic rules, reduces the body-size components to an essentially homogeneous set of particles in relation to trophic uptake processes. While these particles are still identifiable as system components, they can nevertheless be assigned definite roles in the trophic exchanges according to the defined *KFR* parameter space; hence their ecological interaction can be measured in the spectrum as an orderly and predictable syntactic transmission of energy from one group of particles to another.

The earlier exposition of the factors that affect middle-number systems implied that understanding how they operate and on what basis they can be controlled depends on our ability to correctly define the scale of operation of the parameters in relation to outputs. It is evident that a model of transmission of genetic information through a fish population would only incidentally contribute to an understanding of trophic energy transfers. Early studies in fisheries dynamics and management, recognizing that species differ in some of their more important life history dynamics, formulated their models in terms of individual single species. In this way the constituent biological particles could be considered as sufficiently homogeneous for their behavior to be investigated in relation to the defined parameters of change. Such models, like the unlimited energy models of Wilson and associates, did not incorporate any of the elements of interaction of heterogeneous elements that are implied in ecosystem interactive models. But as it turned out, neither have they been able to predict real fish community behavior as fishing intensity has changed.

Ecological models imply interactions among heterogeneous components, which must be defined or aggregated in quite different ways from the single-species models; that is, they must be organized according to a set of rules that defines their relationships with the output production processes. Evidently a biological model composed to describe the body-size distribution of the biomass density, rather than the species composition, meets the requirement. The generality of the biomass spectrum model implies that production output and distribution are based on trophic feeding linkages that operate among different species in a common manner, and are not primarily dependent on particular species definitions. The results are therefore highly variable in species terms.

Knowledge of different levels and types of system behaviors thus enables us to understand the two-natured response of a system according to the degree and level of aggregation of the constituent particles. In the biomass spectrum model, the particles are elementary sets of body sizes with explicit rules of relationship. These enable us to define singular behaviors in relation to the mechanism of energy uptake. In species models, however, we have at present no way to identify a simple set of rules of transformation by which the interaction of information implied in the component species groups can be related to the energetic mechanisms of production system change. Such components are, therefore, still unpredictable on the basis of trophic models underlying the emergent biomass density spectrum.

These differences in species and total biomass models are an accurate reflection of the present state of knowledge of ecosystem dynamics. Evidently an interactive trophic system model is not capable of predicting relative species compositions because species occurrences are not primarily controlled by trophic pathways. In fact, it would appear in general that ecological production models are incapable of predicting particular changes in species composition. The virtue of the biomass spectrum representation in this situation is that it allows us to recognize that the observed irregular behavior of species is the expected outcome of interaction, and so permits us to understand why this should lead to the seemingly unpredictable changes in species abundances. On the other hand, ecological knowledge formulated in terms of a hierarchically arranged energy flow system appears to be the only basis on which we are currently able to measure the effects of disturbance on production, or to assess the possibilities of reasserting control over trends in systems. The way in which we assemble the data on changes critically affects both our understanding and the manner in which we can marshal the forces at our command to achieve particular, and desired, ecosystem states in the face of external economic forces.

The converse situation is also true: a model of transmission of genetic characteristics among components of a metapopulation offers little insight into particular changes in the level of production. However, an important corollary emerges from considering the total evidence on changes in systems compiled in different ways. Namely, that it *cannot* be concluded that species characteristics and production processes are unrelated simply because we can predict them only from models compiled on the basis of different mechanisms. This would be patently naive, but is nevertheless a trap for the unwary, hence worth considering in more detail.

It is true that the biomass structure of aquatic communities probably cannot be predicted from other than trophic transfer. But it is also highly likely that particular features of species affect the *KFR* parameter spaces, defined as the basis for particular configurations of predator-prey interactions, and bear on the outcome of species interactions. For example, it has frequently been pointed out that the higher mortality rates imposed by increased levels of exploitation of the stocks affect the average body sizes of all the species in the exploited community. Such distinctions take us back to the earlier work of Nikolsky (1963), Margalef (1968), and of Ware (1980), all of whom recognized types of population dynamics that may correspond to the distinctly different ways in which species adapt to external forces. They drew particular attention to the physiological processes that underlie particular reproduction and mortality schedules and thus determine the rate of response of the species population dynamics to external forces. A somewhat parallel approach is adopted by Apollonio (in preparation) in his discussion of Gulf of Maine fisheries. Ware (1980), in an analysis that deserves far more attention than it has so far received, related these differences in response by various species specifically to their body size in an allometric formulation.

There is, thus, the distinct possibility that particular styles or patterns of species physiology and behavior emerge at different scales in relation to ecological constraints. The body-size-production relation might already be offering us information that links these two aspects of organism interaction. Where there are important reasons for considering the joint actions of various criteria from two different points of view, such as economic and ecological considerations, it must be remembered that the framing of models is a human intellectual and experiential activity that can be adapted to both the needs and the available information. The density adjustments we have identified as responsible for both the ecological biomass domes and the gaps between subdomes of the spectrum are certainly not independent of the species involved. Thus, investigation of ecological processes in relation to economic forces may well be not only practically possible but potentially a fruitful theoretical field for study. The choice requires only the will and the financial support to undertake it, and the results offer a prospect of simplifying or reducing large areas of the unpredictability we currently face.


## The Effects of Fishing

Fisheries management agencies traditionally focus on controlling fishing mortality, estimated for the identified unit stocks (Gulland 1983; Hilborn and Walters 1992). This is based on the belief that mortality caused by exploitation of particular stocks is the primary variable in both production and yield, and the only one that is within direct management control. While the principles on which this belief is based are clear and logical, the requirement—made evident by the existence of the biomass spectral system and its emergent properties—for dividing the affective forces in a hierarchical system into syntactic and semantic properties at various levels of interaction, and the related question of how mortalities by species are related to community production, obscure the meaning of the traditional calculations. The attendant uncertainties lead to questions about the effectiveness of regulations based on traditional methods.

Our review of system properties makes it clear that identifying both the manner and the scale of action of any cause is of the first importance in understanding its effects, just as it was in explaining the effects of processes controlling the sandpile. Major problems in management may well have devolved from the failure to take these different properties into account. The complex nature of the fishing mortality, and its relation to other parameters, needs to be established in order to clarify our expectations for the effects of any regulative measure on a complex system (Hilborn and Walters 1992).

## The Traditional View of Fishing Mortality and Some Additions

The mathematical formulation of the concept of fishing mortality and its measurement was originally set out by Baranov (1918). It, and its later developments, are described and discussed in detail by Ricker (1975) in a guidebook of computations that is still a basic reference for biologists concerned with management questions. These original compilations specify the assumptions upon which the theory and proposed methods of calculation of species dynamics were to be based. As is so often the case, however, subsequent applications of the methods to the calculation of stock abundance and rates of exploitation became less and less concerned with the possible limitations of the methods due to the lack of agreement between the underlying principles and reality; and became more and more concerned with examining ways to elicit information about the nature of real systems through comparative study of how the results varied and deviated from expectations.

For example, in connection with the assembly of information through which the parameters of fishing mortality are to be measured, Ricker pointed to the central importance of the assumption that the fishing effort be randomly distributed over homogeneously distributed stock elements. Under this assumption, the rate of fishing mortality is defined as $F = qf$, where $f$ is a measure of the number of standard units of fishing effort applied to the fished stock, and $q$ is the constant of proportionality measuring the probability that the application of a given standard unit of effort removes a specified fraction of the population abundance. That is, effort must be randomly applied to the population abundance, which is assumed to redistribute itself following the extraction of a given catch. In system terminology, fishing mortality is treated as a syntactic factor acting uniformly on the stock throughout its range, usually assumed to be a statistical district. Of course, it has long been known that fish population density varies in response to favored feeding habitat, migration patterns, and spawning aggregations, and for other similar reasons that affect the distribution and success of fishing effort (Gulland 1955; Paloheimo and Dickie 1964). Thus, even in these earlier years it was recognized that the strict condition of uniformity would not be met. Many of the early applications of the methodologies were, therefore, intended to test the statistical reliability of the implied basic assumption of a constant relation between "catch-per-unit-effort" as an index of "average density" and the abundance of the exploited stock; or to study spatial distribution of the stock in order to find ways of adjusting for the observed departures.

Early applications of the methodology exposed an important additional practical consideration: that densities of fish stocks are naturally organized differentially by body size, area, and season, and these relationships in their turn determine the distribution and impact of the fishing effort. The existence of these and similar complications was seen in some of the results of the first major application of an international management scheme to the fishery for haddock on Georges Bank in the 1950s. Production in this fishery was considered to be in equilibrium with the effort, with the degree of fishing determined largely by the overall abundance of haddock. The problem that management wished to address was that the fish were captured with a small mesh size that led to high discarding of small, unmarketable, but dead haddock at sea (Graham 1952). This international fishery became subject to regulation under the International Commission for Northwest Atlantic Fisheries (ICNAF), newly formed in 1951. Calculations made in a simple scientific model of the dynamics of the haddock stock strongly suggested that to achieve a higher yield from the natural stocks (expressed as yield per recruit), there should be a delay of approximately 1 to 1.5 years in the average age of first capture. As a result of experiments with different mesh sizes of otter trawls on the size- and age-composition of the catches (Clark 1952), it was decided to require this large, international, off-shore, marine fishery (annual landings of over 50,000T) to change its massive otter-trawl nets from the existing industry standard of a little less than 3-inch mesh to a significantly larger size—specified as 4½-inch mesh.

The scientific spirit of inquiry that accompanied the development of this regulation by ICNAF also led to a determination to regard it as an experiment in which effects anticipated by the theory could be directly compared with empirical results. Accordingly, a special "study fleet" of vessels was permitted to continue fishing with the old mesh. It was believed that this would enable the management agency to trace the development of the course of the landings through an expected initial drop of the order of 10%, as the larger mesh size eliminated the smaller sizes of haddock from the catch, through to a gradual recovery as the saved year-class grew into new larger body-size ranges before being captured. The simple calculations had suggested that it would not be unreasonable to expect gains of up to 25% in returns of haddock from the same basic level of natural production.

Within the first year of instituting the new regulation, however, it became apparent that measuring the effect of the regulation was more complex than accounted for in the simple model. It was expected that the primary effects of the new mesh would be to eliminate the discarding of small haddock. This was indeed established by sea sampling soon after institution of the regulation. It was also anticipated that there might be an increase in the efficiency of the new nets; the larger mesh offered less resistance to the water, hence traveled faster, possibly retaining a higher fraction of the fish encountered. In mortality terms, the regulation could also affect the catchability coefficient, $q$. In addition, however, as the results were examined it became apparent that fishing effort distributions were more strongly associated with local spatial distributions of the densities of the various age or size-components of the stock than they were with the stock's general abundance. In effect, variations in the relative strength of the incoming year-classes of small haddock, and on the period of the growing season in which they were be-

ing fished, meant that use of the new mesh size affected the distribution
of the fleet between Georges Bank and nearby banks on the Scotian
Shelf, and appeared to also affect the distribution of fishing on the bank
itself (J. R. Clark, personal communication, May 1954; see also Colton
1955). In the first year of observation, a number of fishermen shifted
operations to other banks, so that the total landings for Georges Bank
dropped by 15%; but this was attributed to the redistribution of fishing
effort rather than to the releasing of small fish, which seemed to have
caused a drop of only about 4% (Graham 1954). At the same time,
fishing with the new mesh gave advantages in certain seasons and areas,
so that fishermen tended to concentrate on them. In the long term, effort with new mesh responded to the higher biomass densities of the
larger sizes where vessels could fish them, making comparisons with the
study fleet questionable. In the final analysis, it was concluded that
the combination of changes in efficiency, elimination of small fish from
catches, and changing densities at which fish were captured meant that
haddock were actually taken from the bank at larger average sizes than
in former years, despite the fact that the growth rates had not changed.
The final yield advantage in terms of increased yield-per-recruit was estimated to be over 40% (Graham 1957), nearly double that anticipated
from the model calculations.

   This experiment in regulation was counted a success on several
grounds. The mesh-size increase was fully accepted by the fishery, discarding of haddock at sea was virtually eliminated, and the significant
delay in age at first capture had apparently led to a large increase in yield.
In our terms, however, an equally important result was the demonstration that catch-per-unit-effort was primarily a function of the highly
heterogeneous distribution of density components of the haddock population of the area. The new yield was, thus, strongly linked with the
differential distribution of the size components of the various stocks.
The proposed measurement of effects through the study fleet had been
predicated on the concept of an average population abundance, randomly fished by a widely distributed fishery, a truly syntactic concept.
The deviation of the actual result from expectations depended on the
unmeasured, local, semantic aspects of interaction of the fishery with
the distribution of the stock components, and the amount of the deviation was itself an index of how far the simple, homogeneous model differed from the real situation.

There was understandable satisfaction in the beneficial result, but significant serendipity in the fact that the result was higher rather than lower than expectations. While it was clearly possible to calculate an overall value of fishing mortality in the system of which the haddock species was a dominant part, it is, in retrospect, also clear that the potential change forecast by the simple model failed to account for about half the actual change. The effects of the regulation, therefore, needed to be assessed at a quite different level of observation and analysis than was afforded by the fishery information system envisioned in the model. There can be no argument that the whole haddock population was subjected to the effects of a level of mortality that at the purely syntactic level could be related to the size of the total fleet. However, the actual result of regulation was dictated by a complex of distributional factors of internal organization of the stocks and the fishing, which we are calling semantic influences on metapopulation production. These latter effects determined an actual outcome of the regulation that could not be measured using the general mortality information available. Nor, indeed, without more specific information on distribution of fish and fishing fleets, is there any assurance about what fraction of the deviation from expectations was actually attributable to particular semantic factors.

The consequences of the natural organization of populations or communities into components have been slow to be recognized and taken into account. More than a decade after the Georges Bank experiment, Skud (1975) showed that similar distribution effects had been at the base of the inability of scientists to agree on what constituted a sufficient measure of the effectiveness of regulation of the commercially important Pacific halibut fishery. The well-known Thompson-Burkenroad controversy over the results of regulation plagued fishery management discussions for many years after the first criticisms by Burkenroad in 1948. His criticisms had been a factor in the determination of ICNAF to provide a scientific test of application of their mesh regulation for haddock, but were not officially recalled in the later attempts to publicly evaluate its success. Skud found that biases in the measurement of fishing mortality, related to the heterogeneous distribution of components of the halibut population and its effects on the distribution of the effort, were sufficient basis for questioning the strong conclusion of the original investigators that excessive fishing had been responsible for the originally observed drop in total landings, and that subsequent restrictions on fishing had been responsible for an apparent increase in the abundance of the stocks. Once again, in our terms, the misjudgment of the effects is attributable to the difference between the calculated syntactic fishing parameter and the semantic level of response within the system.

Part of the delay in using what was known about the local effects of distribution of fishing on the effectiveness of management can be ascribed to the difficulty of assembling information on local scales. In later examination of this continuing question, Winters and Wheeler (1985) were able to show that there were regular contractions in the range, but not the local density, of a number of species as their abundance was reduced; that is, there was a perceptible and apparently stable inverse relation between abundance and the fishing mortality catchability coefficient, $q$. Their conclusion was in keeping with that of MacCall (1976) that there appear to be systematic deviations of the mortality parameters from the original random distribution model. Such deviations might be used in analyses of fishing effects to show the possibility of generalization at these finer levels of observation and discrimination. That is, these research efforts on distributions provide information that could lead to a generalization of the effects of fishing at the semantic level of cause that is different from, and additional to, that deduced from the simple syntactic models.

To date there seems to have been no attempt to verify or make use of the effects of fishing mortality distributions at these more local scales to institute conservation measures, despite the fact, pointed out by Saville (1979), that a most important practical implication is that such effects lead to a progressive overestimation of population sizes. That is, where population abundance contracts in its range without changes in densities in local areas of persistence, there results a progressively worsening underestimation of the mortality rate as fishing reduces the stocks. During the last decade, however, there have been a number of significant attempts to use this concept to reevaluate the effects of fishing in cases where the estimation procedures according to syntactic principle have failed to detect population declines, or have led to failures of the adopted conservation actions to prevent the collapse of commercial fisheries. Thus, Creco and Overholtz (1990) were able to interpret abundance changes of haddock from Georges Bank by reference to models of the degree of aggregation in population distributions. Similarly, Rose and Leggett (1991) and Swain and Sinclair (1994) showed that explanation phrased at the local distribution scales was necessary to understand changes in landings of groundfish from the Grand Banks and Gulf of St. Lawrence, respectively. Brodie, Walsh, and Atkinson (1998) showed the relation between abundance and range in yellowtail flounder. An equally persuasive and thorough investigation was undertaken by Hutchings and Myers (1994) and Hutchings (1996) for the northern cod of the Newfoundland region, following the collapse of the fishery. They were able to show that the arithmetic estimates of average density gave biased estimates of abundance because of the failure to take into account the changed frequencies of distribution of density components by sizes and ages within the entire fish distribution. The biased estimates of abundance were reflected in a previously unappreciated systematic and stable deviation in the rate of decrease in catch-per-unit-effort with population abundance over three decades. It accounted for the failure to appreciate the actual abundance decline in this wide-ranging and commercially most important fishery in Canada before its commercial collapse.

An interesting contrast is provided by the work of Fromentin and associates (1997, 1998). They studied a 71-year-long data set tracing changes in local spatial and temporal distributions of juvenile cod (*Gadus morhua*), pollack (*Pollachius pollachus*), and whiting (*Merlangius merlangus*) in inshore water of the Norwegian Skagerrak coast. Their results clearly show the effects of the two scales of causes that we are calling syntactic and semantic, but also indicate that relationships between abundance and distribution and various biotic and abiotic causes, which might be supposed to have an a priori simplicity, in fact display spatial and temporal interactions that make them difficult to interpret. After dismissing expectations that the effects could be explained by trends in environmental conditions or food supply, Fromentin and associates imply that the strongest long-term cause of the effects they observed might be an unrecorded local increase in rate of fishing linked to a growth in tourism.

In summary, the evidence from these special analyses overwhelmingly indicates that failing to distinguish between the very different syntactic and semantic concepts of the effects of fishing on populations of fishes has led to a failure of management models and their attendant sampling requirements to detect or anticipate the important economic impacts of substantial disruptions in fish production systems. Ironically, we now see, where the actual rate of mortality is most likely to become too high to allow sustained rates of production, the possibility of detecting the magnitude of the effects is lowest. As late as 1992, Hilborn and Walters (533) could still write, "We believe that spatial variation in fishing policies provides perhaps the most powerful and underutilized tool available to fishery managers." The results of these research efforts to reassess the impacts of fishing point to a continuing, serious need to reformulate the traditional models of the actions of fishing on the stocks of fish. Models need to be developed in a manner that will allow their more immediate incorporation into an appropriate scale of management practices and actions, before, rather than after, mistakes occur in the application of theory to observations.

## Some Extensions of the Problem and the Direction of Solutions

In this section we extend our review of the various scales and modes of action of fishing mortality to other phenomena, to further demonstrate the need for reconceptualization inherent in our hierarchical view of the fisheries system. In fact, the effects that are measured as fishing mortality depend on an even more complex array of factors at many scales and levels within the whole system. Among those most obviously external to the biological system are a number of market factors that relate to price, and technological factors that affect the efficiency and selectivity of capture by various fishing gears. Both are motivated by factors outside the biological system and have an impact on fishing, hence on stocks, whose magnitude is quite unpredictable from the biological system parameters themselves. They can, however, change significantly over short periods and have effects on fishing operations that are virtually unmeasurable in "real time." They undoubtedly have important effects on both the density and abundance of the stocks, and so have to be taken into account in assessing the relation of mortality to the catches and landings of the fishing fleets.

For example, the impact of a single fishing operation by an individual vessel in a modern fleet depends on influences from several different economic and social levels external to the biological production system. There are the intentions of the fishing captain, who has particular strategies of searching and capturing, along with skills for making a profit.

His judgments are based on an experience that balances market prices with operating costs in the context of on-the-spot assessment of the range of fish and fish species that are available for capture. The total impact of a fishing fleet depends, in turn, on interactions among individual vessels, mediated by exchanges of information within the fleet. At a larger geographic scale, the fleet interacts by radio with various "shore captains," who may alter the intentions of either the fleet or of individual vessels because of company strategy. In the world as a whole the distribution of fishing fleets among various fishing grounds is above all else an outcome of complex, unknown, and unmeasurable commercial and government business decisions and intentions, some of which have been made years in advance of any operations. Overlain on all of this, necessarily, are the arbitrary and often cynical interventions of governments, in creating unjustified quotas, subsidizing unneeded processing capacity, and imposing similar manipulations as election fevers wax and wane.

While both the individual and aggregate operations of fishing units may have some relation to immediate biological density distributions through local factors of availability and distribution, or to abundance and productivity through the market considerations of companies, the nature of the feedback relation is far more complex than is envisioned in the simple fishing theory of Baranov and Ricker. In this broader distributional sense, fishing is as much a syntactic factor in fish production as is the weather, or the climate. One might even say that at a level of observation within the production system, fishing would be equally as difficult to predict or control in its precise effects. That is, such effects require management actions at a scale comparable to that of the combined fisheries system created by the conjunction of the sociological and ecological subsystems. The enormous growth of financial investment in fishing fleets, advanced fishing technology, new fish products, and new markets have combined to alter the impact of fishing on stocks, and of the possible responses of the stocks themselves, to the point where it is difficult to discern the kinds of effects that were so simply conceived by the early theory, and that are still implicit in simple management models.

Once, individual fishermen using handlines, traps, or nets made short excursions onto the areas of general seasonal occurrence of stocks, and made individual, usually daily, landings. Now, highly efficient draggers, purse-seiners, or long-liners, of many different sizes, subsidized by financial systems in various ways, handle fishing gears that clear areas of many thousands of square meters of either the pelagic or demersal zones of world oceans over periods ranging from hours to days. In many cases the individual fishing units are massive enough to process and store catches for weeks to months, and in many cases operate cooperatively in fleets, controlled by a central business authority. Such massive technology has the power to locate and entirely fish up or destroy the productive capacity of whole components of the ecological community. Where natural environmental factors, such as rough bottom, fast currents, ice cover, or highly contoured terrain once provided natural refuges to stock components in some places or at some seasons of the year, modern technology now affords ready and full access that can sweep clear almost any target area. Ultimately this enormous fishing power depends on the energy subsidy provided by fossil fuel. There is no obvious equivalent to this in natural systems, which are powered by sunlight. Obviously, the conjunction of natural systems with those devised by human interventions leads to some bizarre properties.

This pervasive, syntactic nature of modern fishing was tacitly recognized in worldwide political actions that restricted the open access of fishing fleets to the world oceans through the enactment of a Law of the Sea Convention in the mid-1970s. However, this onetime event failed to prevent the impacts of powerful fleets, whose fishing within the coastal zones of contiguous countries contributed to continuing struggles in the face of stock depletion. Overall fishing pressures have yet to be effectively controlled. There seems to be little doubt of the continuing need for restrictions, and that such overall restrictions on fishing should originate at the very general level of social and economic actions of nations. Clark (1985, 1990) has detailed the reality of feedback relations between, on the one hand, fishing intensities and, on the other, economic motivations that are independent of the nature of the production system itself and involve investment strategies related to long-term profitability in the economic system. He showed how it is economically possible and potentially profitable for a fishery to completely fish out a stock in the short term and to profit in the long term by investing the proceeds. Such powerful forces, completely outside the actions of the ecological production system, clearly demand actions at a comparable level, independent of the social hierarchies that determine management actions. That is, construction of management models also requires information, analyses, and ultimately control of fishing at levels of system integration that are quite outside the purview of natural production systems.

However, if the goal is sustained production from natural resources, application of *only* syntactic principles of interpretation and control within the fishing system must be considered very skeptically. For example, a reduction of overall fishing intensities in a region by half of present values—much as happened in many areas of the northwest Atlantic following initiation of the Law of the Sea Convention in the mid-1970s—might initially appear sufficient to return fisheries exploitation from the doubled levels of impact reached by the 1980s and 1990s, when worldwide fishing pressures seemed to have reached their all-time maximum. However, during this same period, increases in the level of technology—such as echolocation, and improvements of gear allowing operating on formerly inaccessible grounds or in difficult and dangerous seasons—have enhanced the technical efficiency of individual fishing units on particular population components, such as exist seasonally in special spawning or feeding refuges.

While many possibilities for positive change can be envisaged, the scales of impact of fishing vessels and fleets and of movement and distribution of fish are already well known to fishermen, who need to be consulted for the kinds of information that the abstractions of science and of model building may not recognize, or ignore. That is, effective management systems depend not only on control at the syntactic level, but on information about semantic effects that can be known only at very local social and biological levels in systems. Details such as the presence of special spawning or feeding grounds, or of seasonal migration routes, have often been disregarded in the past as being of too local a nature to be significant to the regulation of whole fishery systems. Our hierarchical view of systems opens the way to seeing the shortcomings of such a priori attitudes.

Problems in identifying and assessing the importance of semantic factors at both the technological and biological scales have been a major impediment to reconstructing realistic models. Fortunately, modern methods of analysis can predict the consequences of not taking into account the impact of such factors. This improved information technology opens the way to more effective and comprehensive assessment of production systems through a study of alternative models. Understanding semantic effects in particular can and must be seen to have a significant impact on the effectiveness of the management systems. We amplify these statements in detail in the next two chapters.

## Summary

In this chapter we began to make explicit that in biological systems that act as middle-number systems different types of forces have quite different effects on output. Middle-number systems display significant effects of what Rosen, pointing to the importance in biological systems of the organizational aspects of interactions among heterogeneous components, termed semantic influences. This perception of a semantic type of interaction, showing up primarily at local scales in biological dynamics, is markedly different from the usual view taken from physics of syntactic relations among homogeneous elements. In our view, this difference needs especially to be considered in fisheries ecosystem analysis in relation to three traditional assumptions: (1) that fisheries are exploiting uniform "unit stocks"; (2) that predictions and control of production can be applied to communities of organisms as the sum of effects on individual species; and (3) that the application of fishing effort to the exploitation of fish populations can effectively assume homogeneity in distributions of both the stocks and the units of effort that generate fishing mortality.

These assumptions are reviewed in the light of examples that show the following:

1. The sustained production from a stock must realistically be related to the balance among different, sometimes small, stock units making up metapopulations, rather than to the overrestrictive assumption of uniformity of elements within statistical districts.

2. Production in an ecosystem seems not to depend on the factors that determine the success of individual species, but rather on the adaptability of the entire species complex in taking advantage of the energy supplies made available from the initial inputs of energy from the sun. From this point of view, fluctuations in species abundance are not predictable from general production considerations.

3. The semantic effects of the density distribution of stock elements within a stock or a community on the distribution of predators or fishing boats obviates the expected effects of a relation between abundance and catch. The distribution of elements has as large an effect on the output of the system as do the values of mortality incorporated in conventional representations of syntactic relations in traditional models of fishing systems. The previously little acknowledged semantic effects account for the deviations of actual fisheries landings from those currently predicted to result from management.

From this point of view, the causes of production change implied in the hierarchy of influences calculated from the structures found at different levels of the biomass spectrum are quite different from those deduced from numerical models of species of organisms in ecosystems. It is our conclusion that the additional facts about observed natural systems and the corresponding deductions from systems theory dictate a quite new approach to the problems of sustaining yields by natural resource management.

# CHAPTER 9

# Levels of Ecological System Output

It is evident from the hierarchical dynamics that underlie the biomass density spectrum that the output measured as production can be considered at different levels of generality. In what follows here we discuss three such levels, based on the patterns of organization of the biomass spectrum studied in earlier chapters. Underlying the first or integral level, which yields the most general overall estimate of production in the system, are the compilations of information on biomass density and production of all the individual organisms in the community, reduced to the level of homogeneous particles differing only in body size. We referred to this scaling earlier as the "physiological scale" of the dynamics because it is a calculation related to the concept of a common physiology in the properties of the basic elements.

The second, intermediate level of system organization yields estimates of production that correspond to the aggregations of organisms into the groupings that constitute the major body-size domes in density. It is at this second, or ecological, scaling that the trophic transfer mechanism, on which our models are based, can be seen as the prime causal factor behind the observed domelike patterning of the biomass density distributions with body size. Energy transfers between predator-prey components in the body-size distributions reflect differences in average production efficiencies among predators; so it becomes ecologically as well as practically relevant to calculate community production for the body-size groupings represented in these domes.

The most detailed, tertiary level of calculation, as described earlier, corresponds to production attributable to subdomes within the major domes in density distribution. These domes were predicted by our predator-prey energy transmission theory and identified as seasonally stable component structures by Sprules and Goyke (1994). While it is doubtful that they represent the same species from year to year, in some cases they may correspond to certain types of species with particular capacities for production. Perhaps more important than identifying such subdomes with species would be linking the production estimates that they represent to the types of mechanisms responsible for this level of production.

Calculations of production at each of these three levels of aggregation can be based on each of the scales of aggregation of biomass density by body size, combined with the dynamics given by the corresponding production density spectra. Our interest in this chapter is the relation of these production estimates to the types of factors that give rise to them. We need to understand how the parameters of the spectrum are affected by the factors of change, in order to better specify their implications for a corresponding redevelopment of management measures appropriate to a hierarchical system. These parameters are of special interest to management questions about stability of output in the face of external factors, especially the modes of action and levels of the fishing intensity.

It must be made clear that perceptions of potential output and of the management measures appropriate to a hierarchical biological production system cannot be completely harmonized with earlier views of sustaining or maximizing fisheries yields in homogeneous, simple systems (often implying separate species). A hierarchical system definition has additional implications for the effectiveness of recent developments in indirect regulatory processes such as the control of fishing through economic forces, or restrictions on fishing effort through impositions of catch quotas, as well as implications for types of direct restriction of fishing activity. We will discuss these in more detail in relation to management requirements in chapter 10. Here we need to specify as clearly as possible how the views of system output developed from the biomass spectrum will be different from the results expected from adoption of the system definitions traditional in conventional ecosystem or fishery models.

## Production from the Integral Spectrum

In keeping with the description of systems in chapter 7, the scale of variations in biomass spectrum properties at the most general, physiological level can be regarded as reflecting an integral or first-order causal level of system parameters. At this level of causes, the individual sizes of organisms are treated as essentially homogeneous particles, distinguished only by their metabolism in relation to body size, in keeping with the well-known allometric metabolic rule. This corresponds to the reduced level that Rosen (1991) termed the syntactic level of cause. It should, therefore, be amenable to analysis according to conventional methodologies. Estimating parameters of the integral spectrum at this level was described and discussed in chapters 2 and 3.

The integral estimate of system production per unit area is readily derived from combining the parameters of the integral biomass density spectrum with the parameters estimated from the corresponding production density spectrum. Calculating the allometric value of $P/B$ appropriate to this physiological scale of aggregation of the organisms was discussed in chapters 3 and 5. As noted there, Dickie, Kerr, and Boudreau (1987) calculated an empirical value of the integral slope to be about $-0.18$, based on data of Banse and Mosher (1980), augmented by data from Humphreys (1979), on the distribution of densities of organisms by body size in their natural environments. This corresponds with conclusions of Fry (1971), who also reported weight exponents in mass specific terms of about $-0.18$ from laboratory measurements of fishes. It remains to be established whether an exponent of this magnitude, derived from both laboratory and field data, differs significantly from estimates of $-0.25$ for the exponent of the rate of respiration per unit body weight of wider taxonomic groups of individual organisms by Hemmingsen (1960), Fenchel (1974), and others, on the basis of laboratory physiological studies. To these must be added the estimate of $-0.22$, deduced from early versions of the partial spectrum and specifically used in interpretation by Platt and Denman (1977, 1978).

These characteristic parameters of the whole production system or of any part of it must, of course, be related to particular levels and types of causes. For example, we concluded that the remarkable uniformity in the integral slopes or curvatures of spectra in chapter 2 among several independent ecosystems in different environments must reflect a general, common property, internal to the nature and distributions of the organisms making up aquatic production systems. On the basis of the data used to calculate integral spectra, we have taken this uniformity to reflect the physiological generality of metabolic relations underlying production processes in individual organisms, as did Platt and Denman (1977, 1978). According to the theories of West, Brown, and Enquist (1997), this pattern appears to be so uniform because, at base, it reflects physical properties of energy uptake that limit its distribution within organisms, thereby establishing a common field for thermodynamic interaction of the various metabolic functions of the component organisms. Under normal conditions this property is also reflected in the integral allometric relation of production to body size described in equation 4.6.

On the other hand, the characteristic differences in position or intercept that appear between ecosystems at this most general level must be ascribed to the external syntactic factors that differ between ecosystems. We earlier concluded that external syntactic factors must be related to the conditions of nutrient supply to the oceanographic or freshwater environments in which the particular density distributions of organisms are found. However, as pointed out in chapter 8, at this integral scale, fishing mortality also acts as a syntactic factor—a point of view that is implicit in conventional management models. We should, therefore, be able to discern its effects on both the level and the slope of the overall spectrum, even though fishing is usually selectively directly applied only to the fish dome of the spectrum.

It is significant in this context that both the overall characteristic slopes and positions in several spectra compiled by Sprules and Munawar (1986), based on information on the phytoplankton and zooplankton body sizes of the total communities, showed significant variations between areas. These variations were at least partly a result of the fact that data on the larger, fish body sizes of organisms were missing from the data series. Examples of more complete spectra, including data on the fish body sizes compiled by Boudreau and Dickie (1992), suggested a uniform slope or curvature but different intercepts or positions among seven systems, which ranged in size from small freshwater lakes to the great North Pacific Central Gyre and included several commercially important marine fishery areas bordering the Atlantic Ocean. At this most general level, the biomass spectrum appears to offer an opportunity to distinguish the effects of external factors, which determine the general level of productivity of ecosystems, from internal factors, which work through the organization of the fundamental biological production processes of the constituent particles to confer on it a particular, internal structural pattern. Thus, changes in the oceanographic regime, such as a change in the water temperature or the level of nutrient supply, would change the vertical position of the integral parabola or straight line fitted to the biomass density data, without affecting its slope or curvature.

Subsequently Sprules and Goyke (1994) and Sprules and Stockwell (1995) analyzed complete spectra for three of the Laurentian Great Lakes, based on special sampling for densities of the ranges of organism sizes from plankton to fish. The fittings of their data for whole lakes to the T-D Model II verified that there were low curvatures with little difference in the levels of intercepts between lakes, and confirmed the generality of the low slopes found by Boudreau and Dickie (1992) in the simple linear model they used. Since Sprules and Goyke also demonstrated seasonal stability of tertiary domes, a priori speculations that slope or curvature of the spectrum would be affected by the heavy fishing mortalities that have reduced the commercial fish groups in recent years (Myers, Hutchings, and Barrowman 1996) are still unjustified. In fact, the review in chapter 7 makes clear that we do not yet have the experience with fish data that would allow us to test whether data fitted to the theoretically time-independent predator-prey models might show evidence of significant progressive modal changes of the sort predicted by time-dependent continuous models (Denman, Freeland, and Mackas 1989). Therefore, the statistical variations in the positions of component groups should be studied further and, where appropriate, taken into account. The commonality of slopes and curvatures that has so far appeared among the historical biomass spectral data makes detection and study of perturbations in recent fisheries data of particular practical importance.

At this point in the development of both theory and empirical experience, we can conclude that, for managers, investigation of the spectrum at the integral level is primarily useful in its potential capacity to provide both comparative assessment of the stability of the ecosystem of a given region on the scale of months to years, and an estimate of level of output compared with other ecosystems. That is, a slope or curvature in common with that found in many other ecosystems suggests a system that is responding adaptively to external forces. The resulting intercept or position is then a measure of relative productivity of this system. On the other hand, deviations of the integral slope parameters from values established by comparative ecological study might serve as indicators of system instabilities. If these were established in relation to external, syntactic influences—including fishing influences—the practical usefulness in calculating and comparing observed yields with potential total production would be much enhanced.

It is too early to speculate usefully on the value of intercept or slope information in connection with other specific environmental effects, such as pollution, although Minns and associates (1987) and Moloney, Field, and Lucas (1991) demonstrated that effects of environmental factors on basic production seem to be promulgated throughout the spectrum and that the theory of energy transmission at various levels helps to explain the effects that appear. Platt, Lewis, and Geider (1984) also noted that the passage of energy from primary production through various larger production elements within the ecosystem is reflected in integral spectral slopes, and provides estimates of in situ rates of turnover that are superior to those based on chemical procedures. However, the consequences of these effects to management initiatives cannot be considered in isolation from effects likely to show up at other levels of the production hierarchy. We therefore turn attention to more detailed aspects of the spectra.

## Production at the Intermediate Level

Estimates of production based on the major domes found within the integral spectrum, especially those associated with the fish-size organisms, are closer to the types of information that have heretofore been sought by fisheries managers. Given what the spectrum has revealed of production relations, calculations must be carried out at this level if we are to meet the objective of comparing the expected ecological production of fish with yields realized by fisheries. Furthermore, because this is the level of organization at which the spectrum reveals a tendency toward stability of internal structure and production in relation to external change, an empirical assessment of the state of the biological community under exploitation needs to start with it. Methods of calculation at this level, using the appropriate allometric measure of production within domes, have been detailed by Sprules and Goyke (1994) and

Sprules and Stockwell (1995), who have also verified that the production estimates for the fish domes correspond well with results given by more traditional approaches and can be made much more quickly and cheaply than the more familiar alternatives.

Of more general importance is the fact that at this level of analysis is found the strongest evidence for mechanisms that underlie the different syntactic and semantic types of system response. Variations in production at this level of analysis therefore have the greatest practical significance in reconstructing management models. While there is still too little experience with empirical data to permit us to draw specific conclusions about the meaning of particular values of spectral parameters, enough is known to help in identifying and evaluating the types of underlying factors, and to begin comparative examination of their likely relations to the objectives and methods of management.

In this vein, it was noted in chapters 5 and 6 that, at this intermediate level of analysis, the predator-prey energy transmission equations giving rise to the domelike oscillations of density with body size can be related to the three causal factors underlying the parameters that describe the structure. We recall that the curvature, $c_0$, of the domes, which has sometimes been approximated by a slope in simple linear models, is given in equation 4.16 as

$$c_0 = -\gamma/\log R,$$

where, as discussed in chapters 5 and 6, $\gamma$ consists of two parts, designated $\gamma'$ and $\gamma''$. The measure of curvature is, therefore, influenced by three relationships: $\gamma'$—the allometric relation of body size to metabolism in the natural surroundings, an apparently stable "internal" cause; $\gamma''$—the added effects of the adjustments of predator to prey densities, arising from various influences both internal and external to the system; and $R$—the body-size ratio between predators and prey, which may vary in response to both the internally determined predator requirements and the external conditions governing both predator abundance and prey availability.

Values of the curvature and of the underlying parameters at this scale were calculated and discussed by Sprules and Goyke (1994) and Sprules and Stockwell (1995) in keeping with the theoretical deductions of Thiebaux and Dickie (1993) about how they may be manifested in observations. We note, however, that additional information on variations in spectral parameters, which helps to understand their nature and their scale of variation, is available in three other studies. In what follows we first review the analysis by de Aracama (1992), which describes variations in the body-size distribution of all species of fishes by subareas within the Scotian Shelf. It permits a perception of the spatial dimensions of coherent structure in a geographically large community. We then review the findings of Duplisea and Kerr (1995), who report a comprehensive analysis of annual deviations in spectral parameters for the fish community of the same Scotian Shelf from their 22-year average. Comparisons among the sequence of annual estimates of the parameter of curvature, $\gamma$, and the sequence of deviations in shape in any year from its corresponding curvature provide a new and practical perspective on the range and variability of annual responses to both natural and human-made perturbations. Finally, MacPherson and Gordoa (1996) report results from a comparative study of spectra taken from five different species assemblages having different fertilities and subjected to different fishing intensities on the continental shelf off Namibia. In the past, parameters related to curvature or slope have been regarded as physiological constants having the simplicity of overall slopes in linear models. However, taken together, the results of all the above studies verify that actual variations at the level of domes within the spectrum respond to a more complex reality and have practical implications for management objectives.

We have already pointed out that Sprules and associates have shown that the curvatures of the biomass density domes within two of the three Great Lakes they analyzed supported the principle of similarity among the body-size distributions of phytoplankton, zooplankton, and fish within ecosystems, but were different between systems. Applying the principle of similarity within systems to information on zooplankton from the third lake (Erie), which had a moderately strong dome curvature, resulted in reasonable estimates of total fish production, when compared with known fishery yields from it. However, as pointed out by the authors, with the exception of the large salmonids, which have been introduced into all the lakes and are maintained by hatchery programs, these estimates of production, based on orderly spectra, are derived from different species groups in the various lakes. From this we may conclude that the curvature of the domes expresses some aspects of dynamic relations between the trophic position species complex and the particular environment in the ecosystem sampled.

These results, added to the information from the preliminary data study by Thiebaux and Dickie (1992), confirm the relation of dome curvature to factors both internal and external to the biological properties of the organisms. For example, the relatively flatter and more extensive body-size dome for the fishes of Lake Ontario reflects the obvious feeding success by the stocked salmonids on the smaller forage fishes. In addition, however, this dome must also be a function of the relatively higher basic fertility of this deeper, more homogeneous lake, which can be concluded from the fact that the zooplankton and phytoplankton domes were also broader in this lake. By contrast, the strong curvature for the separate small lake reviewed by Thiebaux and Dickie clearly reflected the lack of opportunity for large predators to exist in it; unfortunately plankton information is not available. In the case of the intermediate degree of curvature for Lake Erie, we infer it must reflect the fact that this "whole-lake" system is effectively three partially separate "sub-basins" in the relatively shallow, total Lake Erie ecosystem. It seems reasonable to conclude that dome curvature at the trophic position level reflects factors that together might be called the "environmental opportunity" offered by ecosystems to the species complexes in them.

The utility of detailed spectral parameters for defining an ecosystem is suggested by the studies of de Aracama (1992) for the Scotian Shelf. This shelf has an area that is about five times that of the smaller of the Great Lakes. De Aracama undertook a detailed spectral analysis of the fish body-size distribution for the shelf from samples collected at various times within a single year. The overall average distribution for the shelf could be compared with subsample averages taken for smaller and smaller subareas within it. The results suggested that there were two main subsystems within the shelf, one to the north and west of the Nova Scotia peninsula in the Bay of Fundy, and the other to the south and east on the open coastal shelf. Of particular interest, however, is a result for the larger southern part. While there appeared to be a characteristic average body-size distribution of density for the area as a whole, compositions calculated for subareas within this whole yielded patterns that were much alike. The effect was not unlike snipping increasingly smaller portions from a hologram: the basic image was always present, but with increasing noise or variance as smaller portions of the whole system were viewed. This result for the relatively large southern shelf area, in parallel with comparable information on the detailed biomass structure from the lakes, seems to confirm system-wide biological interactions in the distinctive environmental opportunities afforded by an ecosystem. The holographic effect suggested a degree of interaction transcending what would be expected of random statistical variation of a pattern of sequential independent predator-prey energy transmissions among species or size distributions within a common geographic border.

A more comprehensive view of the utility of spectral parameter variation is made possible by the 22-year fish-dome series for the Scotian Shelf, studied by Duplisea and Kerr (1995) and Duplisea, Kerr, and Dickie (1997). The average body-size range on the Scotian Shelf (figure 7.5) was nearly double that of the Lake Ontario dome, hence yielded a relatively flatter profile, reflecting the wider range of species sizes present in the relatively richer, open marine systems (Boudreau and Dickie 1992). The variation of the parameters of curvature of the dome and vagaries in its shape showed, however, that the whole observation sequence could be divided into two periods. The earlier 11 years were characterized by a relatively high fishing intensity, and displayed prominent annual changes in both the curvature and the shape of the domes, indexed respectively by changing annual values of the coefficient of curvature, $\gamma$, and by variations in the details of the fit of the annual size distribution to the calculated curvature for the corresponding year. It was clear that this relative instability reflected both lower densities of the larger gadoid species normally present, and short-term fluctuations and sporadic year-by-year occurrences of different, smaller-bodied species.

By contrast, the later 11-year period was subject to distinctly lower (perhaps close to half of former) fishing intensities, which followed institution of the Law of the Sea Convention in 1978. During this period the annual spectra rapidly developed a flatter and more nearly regular curvature, with the reappearance of the larger gadoids, accompanied by a more regular representation of the smaller species. This set of responses is reminiscent of the sequence of size-composition changes noted by Regier (1973) in the Great Lakes, although in reverse, since in the lakes, fishing intensities increased with time. Somewhat comparable sequences also appear from the work of Daan (1980), who summarized the annual proportions of principal species groups of different body sizes in North Sea landings under variable, though generally increasing, fishing intensities. The data newly assembled in chapter 7 from sample survey data for the North Sea and Georges Bank suggest that they might profitably be subjected to a comparable analysis.

The observations summarized by MacPherson and Gordoa (1996), based on sampling of the Benguela Current system off the southern continental shelf and slope of Namibia, seem to comprehend a relatively wide variety of environmental and fishery influences. The samples showed fish populations consisting of about 90 demersal species, with a size range somewhat narrower than that of the Scotian Shelf, perhaps partly because the smallest size classes seem to have eluded capture. Samples were considered to represent 5 species assemblages in distinct but contiguous biogeographic areas. Two of them, in an area of high productivity, were on an upwelling region of the continental shelf, between 100 m and 300 m in depth. Two others, of lower productivity, were on the upper slopes between 300 m and 500 m in depth. A fifth, of low productivity, was designated at depths greater than 500 m. Pairs of areas at similar depths were differentiated into northern and southern parts on the basis that fishing intensities were high on the northern parts, but low on the southern parts where the bottom was less suited to otter trawling. The results do not offer the detail of the Scotian Shelf studies, since, while they are based on spectra that are mostly parabolic in shape, they were analyzed in terms of normalized straight-line slopes, as were the early spectra of Sprules and Munawar (1986). However, analysis of the straight-line slopes from sampling in several seasons over a period of three years allows the new perception that the effects of fishing and productivity may balance one another.

In particular, in this Benguela system, the spectral slope of what was judged to be the most productive and heavily fished area, the northern shelf, was higher and the diversity of fish species lower than elsewhere. However, while the slopes for the remaining areas appeared less steep, they were also more variable, so that no statistically significant difference could be established between areas of equal productivity but different fishing intensity. The deepwater spectral slope, in an area that had both low productivity and low fishing intensity, was distinctive: the normalized population spectrum had a low average slope ($-0.3$ to $-0.5$) and a highly diverse species population dominated by large elasmobranchs. The four shallower areas, dominated by teleost fishes, normally had intermediate to high slopes (−1.3 to −2.5), reflecting relatively higher densities of intermediate-sized fishes, although with low densities of the smallest sizes of fish in the samples. However, there were occasional invasions of the shallower continental slope areas by large elasmobranchs, and the invasions were consistently reflected in an abrupt lowering of the calculated spectral slopes. These results confirm that interactions of the factors underlying production from the banks, the species complexes available, and the rates of fishing all affect the spectral parameters related to slope or curvature.

When these detailed analyses of spectral data are added to the more general observations studied earlier in this synopsis, we are justified in concluding, first, that the relatively large number of ecosystems that can now be partly or completely represented by the intermediate levels of aggregations of their biomass density spectra all show remarkable stability and resilience of form in the face of external influences. As stated earlier, given a chance, ecosystems appear to organize themselves according to characteristic patterns as effectively as do sandpiles. Their internal dynamic forces are robust, and the patterns of density distribution with body size are resistant to external influences. However, detailed examination of sample spectra at this intermediate level of analysis also makes apparent that both external and internal factors cause variability in the parameters that measure the curvature and shape of spectral domes. We need to inquire whether the principles invoked to analyze and understand events at different scales in sandpiles can also be profitably applied to understanding these variations.

## Two Kinds of Causes of Variation at the Intermediate Scale

Although at an intermediate level of aggregation of observations, the characteristic and stable domelike spectral structures at the trophic position level comprehending fishes are clearly features of the ecological system analogous to the stable overall shapes of sandpiles. That is, at this scale, the dome shapes reflect what we are calling syntactic factors affecting the size compositions of fishes. Responses to these factors must depend for the generality of their effects on the common, underlying physiological properties of the organisms within the "fish" group as well as on their densities in relation to energy requirements. The factors that determine this general result must include the level of fertility of the environmental system, which, given a sufficiently wide supply of fish species able to take advantage of the environmental opportunities, provides the context for the predator-prey dynamics of the dome, in balance with the effects of different levels of fishery removals.

However, the theory of dome formation at this intermediate level invokes additional factors of the organization of energy transmission between the predator-prey ecosystem components, much as was deduced about properties of particular regions in the sandpile: these correspond with what Rosen (1991) classed as semantic properties. Thus, while persistence of spectral domes indicates a common response of organisms of the fish trophic position to syntactic, external properties, we may take differences in curvature and shapes of the spectrum between communities and years to reflect local semantic factors that act differently. Apparently these semantic factors primarily influence controls on the organization and balance of species and body-size components in the subsystems. From time to time, these components of the system exhibit abrupt, even chaotic, changes in the relative densities of biomass body-size groups identifiable at this level of analysis. It thus appears that within certain limits, fishing intensity differences at this year-to-year scale of observation act on the production processes of the community in a manner comparable to that of other semantic factors in community organization.

The scale of spectral variations that appeared in analyses of the Scotian Shelf spectral variations suggested a major change between the two halves of the data history, reflecting the effects of key species elements within the producing community. The same may be true in the most productive and heavily fished region of the Namibian demersal system, where the spectral slope was highest and the diversity of species lowest. Thus, while the overall effects of fishing can be described according to syntactic considerations, as they customarily have been in traditional fisheries models, the evidence from the spectrum indicates significant additional effects at the semantic level. The results appear to have an effect comparable to what might be anticipated within a community from changing rates of predation on selected prey.

However, it is apparent that the effect has been promulgated throughout the density distribution interactions among component groups at various body sizes. Such interactions are inexplicable on the basis of simple, single-species production models. Nevertheless, the spectral evidence shows that at higher fishing intensities these effects are of major significance to the whole fish community, causing measurable year-to-year differences in community structure. It follows that, to be effective, fishery regulations would have to recognize and respond to effects at both the syntactic and semantic levels of causes.

In more usual language, this basic inability to predict particular system components, in relation to management's central interest in maintaining sustainable fisheries, seems to be what is intended by such expressions as "upsetting the balance" of ecosystems. As Bak, Tang, and Wisenfeld (1988) point out, using hierarchical models as the basis, it is possible to recognize that there are elements of interaction among the species components of ecosystems that support the total community functioning; however, when interfered with, these elements result in unexpected and unpredictable changes that make the system appear chaotic from an overall dynamic description; they are akin to avalanche events in sandpiles.

From this point of view, the parameters of the spectrum indicate that the Scotian Shelf ecosystem must have passed through an initial threshold of semantic fishery influences prior to the period of observations available to Duplisea and Kerr (1995), but then exhibited a reversal of the effects as the rates of fishing were reduced during the middle of the period covered by the data. These effects were measurable in terms of the annual dome curvatures and shapes, which differed from expectations of the effects of monotonic increases in fishing intensity on individual stocks or species in single-species models. These results alone would appear to justify management if it were to insist on the construction and study of new, interactive, hierarchical ecosystem models, capable of taking biological interactions into account.

In keeping with the sandpile analogy, it is also clear that effects from the semantic influence on components of the system are not predictable from the same information that is used to fit the simple models that describe fishing mortality according to the syntactic mode of relationship. For example, homogeneous, syntactic models contain no basis for recognizing that removal of the larger, gadoid species in the Scotian Shelf system could, in any way, be related to the increased fluctuations in small-bodied species that characterized the earlier period of higher fishing intensity. Nor is it possible in conventional models to make use of information on differences in predator-prey pairs that are from time to time observed to follow changes in community body-size composition as a simple result of the level of fishing (Duplisea, Kerr, and Dickie 1997). Our general knowledge of the workings of biological systems, fortified by this spectral evidence, leads us to conclude that complex effects associated with high local fishing intensities are likely to include changes in the balance among system components (Daan 1980; Fogarty et al. 1991).

Characterization of changes in the Scotian Shelf community at this intermediate level of integration of observations appears to be particularly important to management interests. Here we witness measurable semantic changes in an integral fish species complex that underwent a reversal after the Law of the Sea Convention brought about a temporary reduction in fishing intensity. A similar reversal may have occurred in the changes in species and body sizes that were observed in the Georges Bank data series under similar circumstances. In their comparison of the North Sea and Georges Bank, Pope and associates (1988) found that the slope of the length-frequency distribution of all fish species combined differed between the two systems, apparently reflecting relatively higher fishing intensities in the North Sea. It also appeared that the slope of the numerical distribution of body sizes of aggregated species of the Georges Bank system was somewhat lower after fishing intensities were reduced by introduction of the Law of the Sea Convention. However, unfortunately for further interpretation of the Great Lakes or the Scotian Shelf, we do not have information on spectral parameters for either the Georges Bank or North Sea cases that would permit us to measure and study this evidence of possible differences in what may be an emergent property of aggregated species in ecological systems. In the latter two fisheries, the data indicate a difference from the Scotian Shelf case, in that replacements of densities of various species in relation to reduced fishing were not by the same species originally present. In chapter 7, we discussed these changes in the light of the associated biomass spectra (figures 7.6 and 7.7), as opposed to the length distributions used by Pope and associates, and reached similar conclusions. Detailed information on the effects of fishing on the species complexes in the Namibian system is also unavailable from the analyses undertaken to date, although general effects on spectral parameters, possibly reflecting species changes, seemed to appear.

As was pointed out in chapter 8, in these complex systems, the effects that appear chaotic at the more general level may still be explicable when the appropriate level of cause has been identified and measured. We do not yet have enough of the kinds of information that would enable us to explain the reactions of particular species or species groups to the community events described above. Densities and body size may show characteristic relations to both fishing intensity and levels of fertility, but generalization of the effects can be established only from broader, comparative experience (Myers, Hutchings, and Barrowman 1997). We discuss this question of reversals again in relation to specific fisheries in the next section and in chapter 10. However, if the interactive interpretations by species groups and body sizes should be generally confirmed, the comparative ecology of major fish production systems would offer interpretations of system stability that are obviously important for management. The model structure that can be developed from the biomass spectrum comprehends a wider variety of population effects that are of significance to management. Further study, even of existing data on major fisheries ecosystems, may yield additional results of use in determining what changes in the fish community would be most amenable to deliberate manipulation.

## Production at the Tertiary Scaling of the Spectrum

The Scotian Shelf studies of de Aracama (1992) and of Duplisea and Kerr (1995) make it clear that the subdome structures identified by Sprules and Goyke (1994) are not the only instances of comprehensible structure at the most detailed, tertiary level of analysis in fisheries ecosystems. In none of these data are there indications that a particular subdome can be taken to represent a particular single species in successive years, and there is every reason to suppose that the fish community consists of a flexible and adaptive multispecies organ for transmission of energy in ecosystems. Therefore, inquiry at this most detailed community level is likely to be most fruitful in establishing the apparent ecological equivalence of various species or species groups to the overall productive functioning.

The presence of domes and subdomes with changing body size composition and shapes from time to time can be taken to reflect a strong adaptive capacity of the communities, at least in the more productive regions that have so far been studied. Unfortunately, however, our sparse experience with observations interpreted according to complex models, coupled with the virtual absence of a developed ecological theory of competition, does not give us much confidence in simple explanations for systems in more northern, less fertile regions. Furthermore, because conventional models have failed to provide timely indications of the actual state of the stocks or the impact of the fishery on the production in the communities, extreme control measures have been invoked in reaction at an inevitably late stage. Thus, current experience does not afford us reasonable precedent for interpreting observed effects, or hope for the building of expectations for future stock recovery. Belief in the simple reversibility of the sequences of changes that have occurred in these complex systems is reasonable on the grounds of simple, homogeneous physical models, but is unsubstantiated by evidence in biological systems that likely have a different order of complexity. Since few or no alternatives to outright closure have been discerned, the highest priority should be to collect a broader base of evidence for continued study.

In addition to an apparent reversal of species groups in the Scotian Shelf case, and in addition to the kind of apparent substitution of species by body sizes in the Georges Bank case, short-term simple reversals in abundance of some of the larger exploited species are known to have emerged from the inadvertent fishing "experiments" in the North Sea when two wartime respites from fishing caused sharp reductions in mortality. However, the effects of these fishing intensity reductions were relatively short lived, and the much higher fishing intensities that developed in the 1970s and 1980s dwarfed the levels of change that occurred in the earlier periods. Thus, while we can verify that fisheries ecosystems show a resiliency at low to intermediate levels of exploitation, the relatively high fishing intensities of the Scotian Shelf and of Georges Bank and the North Sea are important in further understanding changes in levels and types of output in response to fishing mortality.

In his review of fisheries trends and possible replacements of species, Daan (1980), in parallel with the interpretations of the Great Lakes by Regier (1973), concluded that the growing intensities of fisheries were responsible for strong downward trends in abundances of the commercially favored larger species in many areas, and that the trends toward increased proportional representation of small-bodied species in the rapid increases in landings in the late 1970s were consistent with the expected effects of increased fishing. A similar interpretation has been invoked by Pauly and associates (1998) for declines in a "mean trophic index" of worldwide fisheries between 1950 and 1994. However, even in the few earlier cases where there appeared to be replacement among pairs of similar species (such as in the pelagic California or Peru Current systems between sardines and anchovies, or in the pelagic Benguela Current system between anchovies and pilchards), evidence for simple replacement was not often unequivocal, and causes often could be assigned either to fishing intensities or environmental conditions (Daan 1980; Kerr and Ryder 1989). In at least some of these situations of supposed balanced pairing, it is now evident that other, additional species were present in the communities and involved, although they are less well known from the earlier data because of inadequate sampling related to the lack of commercial interest in them. This complicates the earlier explanations of change, which often took the form of equation systems describing direct competition between species.

In addition, Skud (1982) showed that in situations of apparent species pairing, the outcome of observed interactions may depend on the degree of dominance of one species over others. Thus a dominant species that shows apparent advantages over others during a period of increased temperature or level of fertility may appear to be adversely affected by the same external changes when it is in a phase of subordination, or is otherwise low in abundance. In Skud's view, the outcome depends on the actual position of a given species in a natural hierarchy of interactions, which in turn seems to depend on relative abundance. Such effects are consistent with what has been called "depensation" in population reproduction. Relevant here are the recent studies of Myers and associates (1995), showing that evidence for depensation within single species is rare among known commercially fished stocks. Although other causes have, of course, been suggested, depensatory effects might, for example, explain the repeated failures to reestablish lake trout in the Great Lakes, and account for the continued absence of some other species from these ecosystems. Only in a few instances are there sufficient data to show evidence for "depensation"; but in these few it appears that the subject species have not recovered from overfishing, despite persistent protection from further exploitation.

The apparent slow response of the Canadian northern cod fisheries ecosystem to fishing closures may parallel disappointments in expectations for early recovery in such important world fisheries as the Peruvian anchoveta, the California and Japanese sardine fisheries, or the Icelandic spring-spawning herring. Beverton (1990), in his review of the evidence related to recovery times for pelagic species, noted that collapses often appeared to be related to delays in detecting reductions in stock abundance, as has been the case with the northern cod. While recovery times in the pelagics were not precisely predictable, in most cases the periods of delay in anticipated recoveries were related to the life cycle length of the affected species. The long life histories of demersal populations are therefore cause for serious concern given the enhanced levels of fishing on all species throughout the world. Cook, Sinclair, and Stefánsson (1997) have warned of danger signs in the cod stocks of the North Sea. While the evidence points to resiliency of the total community in most situations where the evidence is available, intense fishing clearly affects community structures and production at both syntactic and semantic levels, and recovery is not the simple response to reductions in effort implied in purely syntactic models.

Considering the full complexity of causes in ecosystem events, the remarkable stability of the entire fish spectrum in the face of the different environmental and fishery influences, and in the presence of quite different species complexes, encourages us to point to spectral theory as an apt framework for judging the need for different kinds of management strategies. Spectral stability implies an overall balance in energy transmissions, through possibly diverse pathways. This alone is strong incentive and support for reformulating fishery models on principles of energy balance and competition, such as were first examined in a substantive way by the models of Andersen and Ursin (1978) in Denmark. As implied by the interpretations of Skud (1982), such models need to take account of differences in the hierarchical position of particular species in relation to the causes of change. Apparently effects that appear at one level of a hierarchy will imply conclusions that may even be the opposite of those implied by effects seen at another level. Thus, pending the extension of models to include what appears to be the different semantic, organizational effects in the system, it seems we can place little or no reliance on predictions of the productive course of a particular species in a community based on syntactic rules of relationship alone. At present we are unable to predict species production beyond what can be calculated from the growth of the organisms already observed to be present in the ecosystem. In the next chapter we discuss the implications of these observations for the management of fisheries.

## Summary

Our analyses suggest that estimates of production at various levels in the hierarchies of natural production systems may reflect different aspects of their potential productivity, or their liability to external perturbation pressures. The biomass spectral parameters used in this review suggest that all three levels considered may offer significant perspectives on the needs and possibilities for effective management of the affective forces—perspectives that are likely to contrast with the results offered by conventional homogeneous population models.

While integral spectral models have the potential to help assess the adaptive capacity of exploited systems, the variations in the spectral parameters calculated at intermediate levels of spectral analysis offer the greatest possibilities for interpreting the forces and effects of exploitation. A review of calculations made on the populations of fishes on the Scotian Shelf illustrates that the spectrum is sensitive to the apparently compensatory effects that would be expected in energy-limited natural production systems. These effects also demonstrate the applicability of the concepts of syntactic and semantic levels of influence in ecosystems, and direct our attention to the implications of such concepts for assessing the effects of heavy fishing and the possibilities for remedial management. It is our perception that the semantic effects that show up most strongly in heavily exploited populations may be missed or misinterpreted in conventional analyses. Therefore, in assessing modern fishing effects we need to investigate the application of a range of new models able to incorporate a sensitivity of parameters to species interactions.

∾

# CHAPTER 10

# Implications for Fisheries Management

Several conclusions bearing on fisheries management follow from the foregoing chapters. A basic conclusion is that the ecological system represented in the biomass density spectrum exhibits properties that can be related to causes at different levels of integration of the system components. Thus there are distinct advantages to modeling the system's control dynamics in terms of a hierarchy of causes. Studying the ecological system from this point of view makes clear that certain causes of change in system output have different modes of action, depending on their level in the system hierarchy. Furthermore, observed output may be affected by factors at one level but be virtually independent of factors acting at a different level. This combination of effects implies that management actions must be developed and tuned to the specific causative levels and modes behind phenomena in exploited ecosystems, and may call for an appreciably new system of regulation.

In what follows we outline and contrast the nature of the fisheries developed in this book with the expectations of current management models. We use these contrasts to assess the likely effectiveness of control measures that have been adopted in the past, or that seem needed now. We also consider how the new regulations indicated may necessitate the reconfiguration and reorganization of the administration system.

## General Prospects for Management

In the absence of a fishery, the parameter space of an ecosystem, and its consequent dynamics, is driven by short-range interactions intrinsic to its producing elements. In this sense, the system is the joint expression of (1) the syntactic effects dependent on the relation of the common physiological mechanisms of the contained organisms to the general fertility of the system, and (2) the semantic, density-dependent, feedback effects that emerge from interaction of the natural groupings of organisms as system components. The latter organizational mechanisms appear to be driven primarily by predation parameters.

Imposition of a fishery on this system is not a simple matter of an incremental increase of predation mortality, as is generally supposed by conventional fishing theory. Incorporation of external market forces, subsidized by artificial energy inputs in the form of fossil fuel, or ultimately by the availability of money from the economic system, fundamentally changes the system dynamics, resulting in what amounts to a new production system. This new system is a complex, nonlinear consequence of the intersection of two systems. The new system is based on two quite different sets of primary driving forces, where only one previously existed. To the previously independent natural production system is added an economic, technological, and administrative system that has bizarre dynamics compared with the natural system. The combination of effects from the two different motive forces gives rise to changes in both the balance and the types of dynamics involved, and can make it difficult to perceive what management actions are called for.

Under commercial exploitation, the parameter space of an ecosystem, composed as it is of natural elements that have co-evolved with the system, is modified by an enlarged and artificially generated parameter set. In fact, the most general and pervasive changes in the system arise through the strange, extraneous, and powerful extractive forces of external market dynamics, not from the natural production parameters. Some of the properties manifested in this new system—in the form of an increased rate of removals relative to the natural productive forces—can be anticipated by conventional fisheries models. Such "first-order" effects can be measured at the level of syntactic system properties, as in current models. However, fishing effects are primarily expressed in the natural system as disruptions in the organization of predation relations at the semantic level, and these are not taken into account in conventional models.

Stated more specifically, fishing creates a new source of mortality, which unlike the predation mortality parameter, $F$, acts arbitrarily with respect to fish density. In addition, it causes changes in both $F$ and the parameters associated with it, which together would ordinarily control the natural production. On account of both the unusual level of impact and the unusual but selective distribution of fishing effects in relation to various community components, this new source of mortality also translates into associated changes in $R$, the average predator-prey body-size ratio, and particularly into effects on $K$, the efficiency of utilization of energy resources for production. Interactions at the level of system components thus imply that imposition of significant fishing mortality induces different configurations of the parameter space at several levels of the original energy transfer system, all of which will show up as changes in output. These second-order effects may well be as important as the removal factors to the sustained production of the system components; but because they act on the semantics, they cause changes in system behavior that are unanticipated and unmeasured in conventional models.

Also, causes at the syntactic and semantic levels have effects that show up at different scales of observation. As we have pointed out earlier, syntactic effects are conceived to act on the average particle in the average population. By contrast, semantic effects initially appear as changes in distribution of the relative densities of various community components; and because they rarely affect the exploited units of the species uniformly, they are virtually transparent to syntactic analyses seeking information on a change in average mortality rate. In analyses of fisheries according to such models, the semantic effects have usually been interpreted as statistical "noise," interfering with observation of the expected main effects, but otherwise irrelevant to the control system. Ironically, however, these same semantic effects are the first evidence of change seen by fishermen. They perceive the local effects of system disruptions as new and unexpected, hence as alarming changes in catch compared to the patterns familiar from their historical experience. Thus, the two levels of cause affecting the fisheries system—the syntactic and semantic—afford quite different information and give quite different perspectives on the state of the stocks and of the need for regulation at the two system levels: namely, fishermen and management research and administration. The consequences of the legitimately different points of view of these two levels for the effective administration of conservation measures have been vastly underestimated by the level of conventional, centralized management authority as well as by the level of fishermen subject to regulation.

Thus, the actual two-tiered nature of fishing mortality has meant that attempts to control it on the basis of models of simple mortality have been ineffective on several accounts. In the remainder of this chapter we review questions and tentative conclusions about the indicated direction of development for a more effective regulation system. This direction is based on system properties revealed by the biomass spectrum in relation to both past failures to appreciate stock changes and the prospects for predicting future events. Three principal problems arise. The first concerns the importance of interactions of parameters at the semantic and syntactic levels to detect and assess the impact of fishing. Evaluating their importance is directly related to the problem of predicting allowable levels of fishing effort. The second problem concerns the effectiveness of various means of controlling these effects; regulation that is not specifically aimed at the appropriate level and mechanism of cause is unlikely to achieve the desired result. The third problem concerns the nature of an administrative system capable of acquiring and making use of the kinds of information necessary for effective control. The dynamics of the fisheries dictate the need for a multileveled system of both information flow and regulation.

## Interactions of Parameters and the Problem of Setting Limits

Modern intensities of exploitation that supplant the natural semantic predation processes with powerful selective fishing mortality are motivated by external market forces. This removal, therefore, is far from being a component of negative feedback in the energy system. Instead, it acts as an independent syntactic factor, which can lead to a condition of disequilibrium within the natural system. This may well be the most pervasive and damaging property of modern fishing technology. Therefore, our first concern over prospects for management must be that fishing exerts effects on production that, in real time, are as general and independent of the natural production system as are the factors that determine its nutrient supply, but act in the opposite direction. To revisit the sandpile analogy, a significant fishery moves the system from a situation where there has been tinkering with the adhesion properties of the individual grains to a situation commensurate with the application of external forces that modify the sandpile's gravitational field, or even remove the sand!

The replacement or overlay of the community semantics with sweeping syntactic change poses problems in both detecting and predicting the effects. Natural production systems have a built-in resilience to external perturbation or they would not have survived. Empirical evidence of their adaptive response is present everywhere in the continued occurrence of a biomass density spectrum under many different exploitation regimes. However, since modern levels and kinds of fishing are essentially disruptive, at a certain point they will overwhelm the system's adaptive capacity. The extent to which they disrupt and supplant the natural semantics of fish production systems is, thus, the ultimate measure of their impact, and one that needs to be understood and evaluated. It is a matter of practical necessity that administrative systems charged with managing fisheries have the means not only of measuring the total impact in terms of mortality but of relating the impact to the mechanisms through which it acts so that regulation can be directed toward the motive forces responsible for the observed perturbations.

At the very simplest technical level of interaction in data describing the biomass spectrum, we note that the parameters of curvature or slope and vertical position or intercept that appear at the integral level of aggregation of data discussed in chapter 9 are usually treated in analysis as though they are independent variables. This is in keeping with the methods of analytical geometry that underlie mathematical curve fitting. However, we have already noted biological reasons, related to the selectivity of fishing operations, for looking more deeply into any such assumed independence. At the technical level, therefore, we may ask whether the level or average position of the spectrum has a bearing on the curvature, or on the variability of the shape of a dome in relation to this curvature. This is the same as asking in more general language, "Does the level of fertility or productivity in the system affect its resilience to fishing?"

There is a widely held belief that ecosystems with rich species densities have a greater potential for a stabilizing negative feedback response to external perturbations than do depauperate, northern systems of lower annual productivity. In the current environmental vernacular, there are frequent references to the "vulnerability" or "fragility" of northern ecosystems in the face of exploitation or pollution. Somewhat in support of this belief, the historical evidence shows that among exploited aquatic ecosystems, the highly productive communities of Georges Bank or the North Sea have exhibited a stability of spectral properties that indicates a remarkable resilience to relatively high fishing mortalities, at least for some species (Myers, Hutchings, and Barrowman 1996; Cook, Sinclair, and Stefánsson 1997). On the other hand, except for the two or three years just before the final commercial collapse of the cod fishery of the Grand Banks and other northern banks off the northeast coast of Newfoundland, this system did not appear to be subjected to fishing mortalities that were outstandingly higher than in other regions (Myers, Hutchings, and Barrowman 1996). Does this imply that this relatively species poor and less productive region may, in fact, support a system that is less resistant to fishing than ecosystems elsewhere?

This is not to say that the rich fishing grounds of Georges Bank and the North Sea have not undergone significant changes in their species complexes. Nor that other cod stocks, such as those of the North Sea, are not also in danger (Cook, Sinclair, and Stefánsson 1997). However, though spectral parameters have not yet been calculated for these regions, the more productive systems have so far exhibited a sustained biological capacity for continued overall production. This is at least partly dependent on the rich variety of species present. The different species appearing at different times and sequences in the landings in these areas are in contrast with an apparent lack of alternative fishing opportunities on the extensive Grand Banks and Northern Shelf ecosystems as the cod have disappeared. A functional relationship between resilience of the system and the variety of species present in it would likely be reflected in biomass spectrum curvature and positions, but such criteria remain to be established. Even the possibility of indexing them is of great interest to the question of sustainable fisheries. The data to address the question evidently exist in files amassed by such national agencies as the Canadian Department of Fisheries and Oceans, but the requisite analyses have yet to be undertaken.

It has been pointed out by Hutchings and Myers (1994) that before the present situation developed, the northern cod production system in the Newfoundland region seemed to be in balance with the rate of fishing for more than three centuries, with landings frequently in excess of

250,000T between 1870 and 1930. During the earlier period, fishing was mostly in the form of hook and line or trap fishing, pursued by relatively small boats during ice-free seasons. This situation rapidly changed after 1962. With the advent of large otter trawlers with modern technology, the catch briefly exceeded 800,000T, and the rate of fishing increased beyond the reproductive power of the principal cod stocks to sustain themselves. Despite the brief respite following institution of the Law of the Sea Convention in 1978, the cod stocks rapidly collapsed again after 1988 in the face of the steadily increased fishing effort. In the second case equipment was used that had a capacity to operate through the ice, hence in areas that may formerly have been seasonal refuges for certain components of the species stocks. The additional question this poses is whether the deleterious effects of fishing were due simply to the higher rates of removal in the syntactic mortality sense, or whether there is a need to separately or additionally identify semantic disruptions of the production processes.

The series of studies by Hutchings, Meyers, and associates makes it clear that the effects of fishing were also evident in the declines in abundance of spawners. In fact, measurement of the overall abundance of the spawning component of stocks, toward which fisheries research has been working for many years, can be seen as a major advance in understanding. For example, Cook, Sinclair, and Stefánsson (1997) calculated the functional overall relationship between spawning stock, fishing mortality, recruitment, and yield for the cod in the North Sea, showing the essentially curvilinear nature of the relationships. As they point out, the management implication of curvilinearity in these first-order effects is that an attempt to hold fishing mortality in the region of the maximum is an extremely risky undertaking. Expanding these considerations to the whole of the natural fish community may, in fact, cast some light on the reasons for abrupt species changes in heavily fished multispecies systems. There is a direct inference that declines in recruitment are specifically related to decreases in spawners because of high fishing mortality. However, the implications of earlier studies of the effects of regulations on fisheries, such as the Georges Bank haddock or the Pacific halibut fisheries, are that, while all these first-order effects indicate the direction of needed restraint, the essential knowledge of the mechanisms through which they act is not revealed by analyses at these syntactic levels, nor is the impact of fishing accurately measured without information at the more detailed distributional level of the community components.

In the case of the northern cod stocks, Hutchings (1996) showed the remarkable changes in the density distributions that accompanied the stock declines. Clearly, semantic effects were as evident in these stocks as they have been elsewhere. That is, northern cod stocks have undergone changes in the organization and distribution of stock components to at least the same extent and in the same manner as did the Georges Bank haddock or the Pacific halibut. In these latter cases, the aggregated distribution of stock and fishing components meant that the analysis of first-order effects of fishing using the conventional, homogeneous, syntactic models failed to measure the real impact of regulations. In each case, the second-order, semantic changes in distribution alone have been sufficient to account for the difference between the effects predicted and the results observed. And, of course, distribution changes are only one aspect of semantic effects.

A parallel situation seems to have occurred in collapses early in the exploitation history of fisheries for sturgeon, lake trout, lake herring, whitefish, blue pike, and other species in the Great Lakes ecosystem. Lakes are generally regarded as less fertile than marine systems (Leach et al. 1987), and these larger-bodied species that were initially taken in the commercial yields were early casualties, disappearing successively from the catches soon after the fishery targeted them. While much of the early evidence concerning them is too general and anecdotal for retrospective analysis, the course of exploitation effects in these important fisheries appears to be no different than it has been in the major marine ecosystems. For some of these species, there is little doubt that other factors, predation by sea lamprey in particular, had important effects, but clearly not in all instances. Fishing, in most instances, appears to have had the major impact. Continued failures of management to anticipate or control changes in species composition in fisheries throughout the world indicate that broader avenues of investigation are needed.

Comparing semantic information, which reflects second-order effects, with the syntactical information currently relied on in fisheries analyses, the question becomes, what are significant changes? In these situations, the application of biomass spectrum analysis to the Great Lakes ecosystems by Sprules and associates has opened new possibilities for a sound comparative ecology based on the principles of energy balance and similarity between systems. Their use of data from appropriately designed sampling programs clearly indicates the feasibility of a data compilation that can comprehend the broader analyses of both first- and second-order effects that are indicated by fisheries data in general. In addition, the empirical analyses of Duplisea and Kerr (1995) used survey data for the Scotian Shelf demersal fisheries to study the semantic properties of fishing in detail. The vicissitudes they detected in parameters over time, like the first-order effects predicted by Cook, Sinclair, and Stefánsson (1997), may not have been monotonic functions of fishing intensity. That is, the erratic behavior of the shape parameters of the biomass spectrum at high levels of fishing stabilized abruptly following the imposition of the Law of the Sea Convention and the drop in the rate of fishing that followed from it.

Evidence from MacPherson and Gordoa (1996) of simultaneous changes in slope of the spectrum and diversity of the exploited species complex in the Benguela Current system is further indication of what empirical study offers at this level of observation. Such effects clearly need to be checked in other fisheries, for many of which comparable survey data already exist. But even without further data analysis, it is clear that in combination with methods of analysis at the syntactic level, assessment of effects at the semantic level provides more sensitive, local indices of system conditions, which seem likely to help evaluate at least those indices of vulnerability that have been correctly recognized by fishermen as warning signs of ecosystem stresses and incipient changes in output. To accomplish this, the assessment models used by management to interpret fishing effects need to heed the sometimes simple orders of effects that have been adequately measured (Ludwig and Walters 1985), while at the same time making provision for a complexity capable of accommodating the actual observed natural system effects.

This is not to say that the process of fishery analysis and regulation can be based primarily on empirically established relations. The likely impacts of various causative factors need to be extensively explored in model studies if we hope to predict levels of permissible fishery removal more precisely than can now be done. An initiative in this direction took the form of international studies of the possible need for more stringent fisheries regulation. These explored the relation between maximum allowable catches determined from the summations of results given by individual species models, and an overall yield index for the system compiled from empirical data on total system yield and effort and using equations developed from the Schaefer (1967) model. The results of these studies by Hennemuth and colleagues have been summarized by Brown and associates (1976) and Sissenwine and associates (1982). The studies showed the difficulties of building complex models when one started from a knowledge of the details of exploited species, but lacked the supportive force of underlying energy balance theory, which might now be provided by incorporating the biomass spectrum. Still, there was a tacit understanding that the more conservative approach to management indicated by the results of community-level analyses deserved careful attention. Such studies—in conjunction with earlier attempts to model energy budgets for ecosystems (Steele 1974; Mills and Fournier 1979; Cohen et al. 1982), or recent studies of the Steele and Henderson (1984) modification of the Schaefer models to consider predation effects (Spencer and Collie 1997)—demonstrate that a knowledge of the system does indeed result in an estimate of output that is different from that derived from a summation of its single species parts. Whole system models that differentiate between levels of cause in the hierarchy of system relationships are a place to begin examining the value of complex, multispecies model development.

At present, the likelihood of interactions among parameters of production systems in the presence of heavy fishing pressures means that a study of the empirical evidence is an essential element of research oriented toward providing improved models for management concerns (Collie and Spencer 1994; Spencer and Collie 1997). It is now possible to conclude that the syntactic level evidence currently available for assessing the state of fisheries systems provides only a minimum estimate of the likely impact of fishing mortality on the natural productivity of stocks. Given the recognized imprecision of this methodology and resulting delays in predictions, management agencies perhaps have had few alternatives to the drastic actions of regional closures, such as were invoked for northern cod off Newfoundland. However, with the methods of analysis made possible through spectral parameters, in association with the wealth of survey data that has been assembled (but little used) on the compositions and distributions of stocks of the entire fish community, significant improvements can be made in identifying causes and effects. These should allow blanket restrictions on landings to be supplemented by more specific measures.

## Methods of Controlling Fishing Mortality

An important management concern must be the effectiveness with which fishing effort and its "mortality" effects can be controlled. The recent history of management shows a frightening record of failures and clearly indicates that recent, newly high levels of intensity have given rise to unanticipated effects. Some of these—such as the obvious lack of appreciation of the fact that a constant fleet size leads to rapid increases in mortality rate because stock distributions contract as abundance falls—are clearly ascribable to the use of simple models, which have depended on observational data that is far removed from the requirements of appropriately complex models. The belated reactions when overfishing becomes obvious have understandably tended toward the severe. But the economic and social devastation that results from blanket closures is a sobering index of the seriousness of the failures—closure of the Newfoundland cod fishery in 1992, for example, put some 40,000 people out of work in an area where there are few alternative ways of making a living. More realistic alternatives are needed on a virtually emergency basis.

Soon after fishing fleets began their expansion in the 1950s, international fisheries organizations worldwide attempted to control mortality rates through restrictions on landings by species and areas. The original basis for this method was the belief that control of dockside landings was the most direct and practical approach, as opposed to control of catch at sea, where effective enforcement was deemed to cost more than most countries were prepared to pay. Control of landings was believed likely to be effective, and seemed the best possible negotiating route to the control of fishing effort in an international forum where support of the building and deployment of vessels was a closely guarded privilege of national governments with sometimes very different economic systems and policies. The failure of this method to effectively limit the steadily increasing amounts of fishing eventually led to adoption of special fishing regulations in the Law of the Sea Convention in the mid-1970s, and reformulation of existing international fisheries commissions.

The intention of this renewed attempt at regulation was to reduce the apparent trends toward increased biological "overfishing" in international waters by placing both the responsibility and authority for deployment of a sustainable level of fishing effort in the hands of single coastal states, thereby avoiding lengthy debates and some of the problems of delayed application of needed restrictions. As can be seen in the records reviewed here, the Law of the Sea Convention regulations caused an initial significant reduction in fishing, which lasted for as long as five years in some of the main fishing areas around the North Atlantic (partly because the regulations diverted the distant water fleets of several nations to other areas). However, political and economic motivations within and between countries soon allowed a resurgence of fleet activities in the most intensely fished areas, even in the face of steadily falling total catches from them. The continued record of collapse, and of apprehension about the state of remaining fisheries, clearly shows that basic problems were not solved by creation of this new authority.

One of the first postulates of the theory of fisheries management has been that the failure to stem the trend of increasing effort results primarily from a simple, addressable fact. This postulate was first applied to fisheries by Gordon (1954), who pointed out that success in management is tied to some means of overcoming the economically and biologically unsustainable nature of a common property right under open competition. The means indicated was either imposition of a central authority, or a market-driven self-motivation through the creation of property rights to fish. A second, corollary postulate, pointed out by Ashby (1964), is that hierarchical systems are ultimately uncontrollable by centralized authority, because centralized authority commonly lacks the "requisite variety" to counter the varied actions that can be taken by lower levels in the hierarchy to avoid regulations and penalties. That is, central authority cannot effectively control a system unless the governed are in essential agreement with the objectives and methods of the governor; even tyrannical systems ultimately collapse because, with time, they fail to keep the support of a sufficiently broad segment of the governed. Thus, assignment of authority to a single country would have to be explicitly reinforced from time to time at various levels of the administrative hierarchy. The added problem of maintaining good relations between authority and the fishermen it governs is one to which a lasting solution has yet to be found in fisheries.

Both postulates of the conditions for successful governing are applicable to the current situation, where, in addition to the stubbornly persistent failures to stem the trends toward overfishing and stock collapse, there is increasingly serious erosion of both the information collection and application phases of the administration of proposed regulations. We are now proposing an additional postulate, which is that the basic level of information collection and interpretation must correspond with the real level of complexity of the system governed; because, to be effective, regulation must be applied to mechanisms that exist only at specific levels of system behavior.

This book is not the place for detailed discussion of alternative regulation measures; the essential relevant point about the biomass spectral system of energy balance relates to its different perception of the factors that sustain production in natural production systems and its consequences for other conservation measures. Specifically, spectral theory implies that the present basic premise of regulation of catch by individual species cannot sustain multispecies fisheries, because system production is not an additive function of individual species. That is not to say that an administration cannot apply rules to individual species; but the means of arriving at an allowable level of effort, and a corresponding allowable catch, cannot be based on the summed estimates of individual exploited species given that the control of production is determined in the system at the level of total amount and distribution of the total energy flow. That is, if production systems are organized according to biomass density functions, irrespective of species composition, then it is folly, or at best irresponsible, for management agencies to think that single-species management is the appropriate choice for them. Clearly it is not.

Effective control of exploitation must be based on the various mechanisms that motivate the dynamics of the system in relation to its total energetics. These include the natural control in the total nutrient energy balance, in relation to a yield that might then be apportioned according to the economic desire for particular species. But effective controls are even more importantly associated with the adaptability of the whole biological system to the important disturbances of its semantic organization. Such disturbances are made possible by an external energy supply in the form of subsidies to fishing industries throughout the world. These basic facts of system configuration and motivation call for a fundamental change in the management approach. Management cannot start from a position that looks first at the individual, economic, species outputs. The truism that a system is more than the sum of its parts is nowhere more evident than in the need for management to look at the whole biological organization of the industry output in relation to the kinds, amounts, and distribution of effort on the system components provoked by external market forces. We need to consider the important consequences of this different view.

As was implied at the beginning of this chapter, the first requirement of management is to assess the total state of the production system, in relation to the various demands made on it. Indices previously available, such as the basic primary, phytoplankton production, or even production in compartmental food webs, have clearly been too general to be practically useful. The niceties of judgment are improved by the more precise measurement of the parameters that has emerged from the functional formulation of the biomass density spectrum. However, at this general level, the system production information is still of limited use. A second, more timely requirement is an estimate of the total, syntactic level of natural production within the exploited domes, compared with the actual yields from them. This may also be calculated from the spectrum, and may include more than the fish dome. Third, and in some ways most importantly, must come estimation of the tertiary, semantic level effects that appear among the totality of the natural stock components within the domes of the spectrum, together with the fishing effort distributions and intensities associated with them. The analyses of Duplisea and Kerr showed that this can be indexed by the spectral parameters; however, there is a place for more comprehensive information on distributions of stock components, such as the spatial concentrations of spawners, and a need for special models that can examine such effects (Spencer and Collie 1997). Only with such information in hand can estimates begin to be made of the fraction of the total production of any particular community component of any species in any particular region that must be allowed to survive in the interests of a sustained, exploitable biological system. This might seem a long list of requirements. But given fish distribution surveys, in fact it can be met more easily and quickly than the present requirements of age-structured models, which are used to estimate syntactic levels of fishing mortality and spawning stock size annually by species and area. This latter information on age-class abundance by stocks is inevitably delayed by the need for data from the sampling of commercial catches.

However, these considerations imply that certain strategies invoked by present approaches to management need to be replaced. If the current dependence on species-specific fishing rights, exercisable through licensing or through the auctioning of rights to fish (individual transferable quotas, or ITQs), cannot bring about a new balance between local fishing intensities and production because single species do not control natural production, new principles must be found. In addition, because a particular level of catch is not a reliable measure of mortality rate when mortality rates are high, a quota scheme based on landings requires rapid "adaptive" adjustment as the area of distribution diminishes. This, of course, will always reach a point where the need to change is faster than most catch information can be collected and assimilated. Therefore, some a priori projection of system effects at an appropriate level of detail is necessary, and requires more realism in the conception of the system. In fact, it appears from such problems that indirect control of fishing through dependence on catch limitations may well be an abstraction that cannot, in practice, control the steadily increasing incentives to cheat that will always be motivated by economic forces very little related to the catch in a single region or of a particular type of fish at a given time. Fishermen who start out agreeing with authority to restrict their activities in relation to a particular species or place or time are likely, for a number of the reasons indicated above, sooner or later to become antagonistic to it. Some will begin to exercise their degrees of freedom to circumvent controls, and others will follow. The result is that, in accordance with our third postulate, the authoritative administrative system now in existence almost everywhere in the world will exacerbate interactions of fishing effects on natural production processes, rendering ineffective the administrative actions intended to permit recovery by creation of property rights, simply through the administrative system's failure to differentiate the mechanisms behind various effects.

These problems of ineffective management seem to stem jointly from mistakenly identifying the nature of the impacts of the fishery on its own future catch, and from failing to perceive the different orders of factors that control the natural system production. Both have to be explicitly recognized, but, even more importantly, have to be addressed together by perceiving what regulations are needed as well as the form of the regulations. For example, even with their better understanding of the implications of the semantic level effects that they are the first to perceive, fishermen are powerless to exercise their responsibility to prevent abuse to the natural system as long as they are separated from the authority over it. The implications of this for the administrative system will be discussed in the next section.

In conclusion, we point out here that because fishing effort at the syntactic level of most modern sea fisheries has little or no feedback relation with the biological yield, control of mortality has to be directly applied to the vessels and gear that do the fishing. That is, there seems to be no effective alternative to development of a control system that is completely external to dependence on feedback relations to catch. Indirect control by limits on catch does not guarantee a decrease of mortality. Furthermore, because of the complex nature of fishing mortality, regulation may need to include specification of not only fleet size, but its precise character by area and time. Semantic level effects of fishing strongly indicate that fishing affects fish production by interfering with its internal organization. Up to a point, interference elicits adaptive actions by the community of organisms; so expression of adaptive mechanisms to this point is a cost that must and can be borne by the natural system. However, because there is a limit to system adaptability, there are additional information costs for any attempt to approach that limit. At a minimum, this also means providing detailed information on fishing activities, and on the responses of the natural stock elements, so that, where necessary, local area and seasonal closures can also be imposed in response to local configurations of the natural producing system. In a sense, avoiding the problems of dictatorial control, which become greater and greater as the regulation system has to resort to more and more detail, clearly depends on how that control is administered. Those involved in fishing operations must be willing to share the costs of and take joint responsibility for detecting where regulation and planning are needed and administering the necessary controls.

## Implications for Administration

Perhaps most important for the successful administration of fisheries are the consequences of the many different perceptions of the need for regulation by participants at different levels or positions in the exploitation and resource use system. The fishermen who are supposed to be restrained by regulation have the greatest sensitivity to changes at the local level. They do not have the means to measure syntactic effects like mortality or spawning rates, and thus have a limited ability to deduce the overall significance of what they see. However, because they see the signs of disruption of the natural system, they are the first to detect possible danger signs. By contrast, partly because of the focus on the single-species models intended to measure fishing mortality, management agencies have been guided by a different view. For a number of reasons they have maintained a certain distance from local operations, partly as a traditional tactic of a police force, but partly because they perceived the effects noted by fishermen as largely irrelevant. Ironically, this has meant that by the time management has acknowledged the significance of the biological degradation of the system, it has already lost the confidence of those who must be managed.

But there are also other actors in the system. For example, the "consumer" influence is felt in a number of ways, both through the markets that demand fish for food, and through the various sports, environmentalist, and social action groups, all of whom demand a voice in determining how the fishery should be managed. Myriad legitimate demands are made on fish stocks, which economic theory attempts to deal with under the rubric of various consumer and producer "surpluses" or net social benefits. One of the virtues of the latter perceptions is the recognition that there is a limited system output in which the shares demanded by various groups are inevitably in direct conflict with one another. This is the reality within which the overall management system must be built.

Detailed discussion of the shape of a fisheries administration is no more the business of this book, or the competence of its authors, than is detailed discussion of the methods of control. It is our business and intention only to point to the implications for this complex, multilayered, hierarchical fisheries system of the properties that arise from recognition of its biomass spectral nature. Both the system's characteristics and their scope seem likely to shape the effectiveness of particular patterns in the administrative structure. More specifically, in a system as complex as the one that arises from imposing a fishery on a natural production system, the effects of perturbations in that system caused by exploitation or pollution will always first appear at the lowest, most detailed level of observation. Therefore fishermen, who have the longest and most intimate experience of the system, will be the first to have evidence of change; and often the various naturalist and environmentalist and fishing community groups will be the first nonparticipants to develop an appreciation for fishermen's concerns. It is also at this detailed level that the likely consequences of those changes are the most difficult to discern. A common understanding requires both the systematic compilation of many details, and a careful interpretation that likely involves at least some form of complex models.

Unfortunately, however, there are serious disagreements among the public about appropriate forms of resource management. For instance, there is the contention among various sectors of the sport and commercial fishery and numerous other user groups in the Great Lakes ecosystem. To some users of the resource, the capture of a trophy-size exotic fish species is a much-sought-after experience, even if the fish is inedible because it contains unacceptable concentrations of toxic contaminants. To others, who perhaps remember how such systems once were, fishing for inedible trophies is odious. To an outsider aware of the real dilemmas facing major world fisheries, a dispute between two such user groups may seem amusing or trivial. Nevertheless, such situations represent policy conflicts that plague the current generation of Great Lakes fishery managers. For managers to succumb to the pressure applied by some lobby groups would virtually result in converting the former Great Lakes fisheries into a kind of hobby farm, with hatcheries for exotic species— produced, if possible, at public expense—to satisfy one or another vociferous and sometimes wealthy and politically powerful special interest group.

Although such a scenario is real, and possible, the saving feature for the Great Lakes fishery is the diversity of other user groups, ranging from native peoples to industrial water users, who also have a stake. The cacophony of conflicting demands has been for the best, in the sense that traditional monopolizers of the Great Lakes system can no longer continue unchallenged. In fact, under the aegis of the Great Lakes Fisheries Commission, and the International Joint Commission, rationalization of human use of the ecosystem is perhaps as advanced as anywhere in the world (Kerr and Ryder 1997).

Although past models of fisheries have largely oversimplified the study of the fisheries ecosystem, this can be overcome by appropriate action of fisheries administration. That is, once the adaptive properties of model building and its relation to regulation, long ago espoused by Walters and his associates (Walters 1986), have been understood, the abstract concerns of the first-order, syntactic level of effects, which up to now have not been related to the effects observed by fishermen, can be juxtaposed with fishermen's semantic level concerns. These concerns have been developed over many years of first-hand experience by those with a comprehensive local knowledge. Our previous inability to understand and incorporate these concerns into the system of interpretation is, therefore, one vital aspect of the present failure of the management system itself. Bridging this first dichotomy in the information collection and interpretation system appears to call for a much more local level of interaction of these aspects than heretofore possible under a fisheries administration system that has seen itself as lawgiver and law enforcer for the system.

Clearly the larger system framework for fisheries dynamics entails a major problem in establishing an effective hierarchy of responsibility and authority that can interact to formulate and carry out the necessary regulations. Here again, the biomass spectrum properties are important, in that they point to interaction in the total energy system as a necessary instrument for both interpretation of the needs and implementation of the remedies. That is, no single segment of the fishery can dominate the administration of regulations, since the objects of exploitation within the system are subject to rapid and sometimes drastic change, which (as was pointed out earlier) even appears to be chaotic when the appropriate level of information is either not available or not utilized. Thus, for example, assigning responsibility for the fisheries of the continental coastal shelf areas to a single authority through the Law of the Sea Convention has not succeeded in protecting the stocks. Regulation of fisheries by a single country apparently does not effectively guarantee the beneficial properties of a sole owner able to protect his investment, as was envisioned by Gordon (1954). The failure cannot be simply attributed to a lack of information, even if ignorance of the real nature of system organization has been a factor.

In the practical world created by the confluence of biology, economics, and sociology, societies everywhere have had to resort to government as a means of balancing the different and conflicting needs of the many different interest groups. As pointed out by Wilson and Dickie (1995), the many different levels at which the demands in fisheries are expressed have the same sometimes bewildering complexities that appear in other areas of society. Traditionally the fisherman has been represented as the central figure. But with the growth of the industry and the technological society of which fisheries are a part, even that traditional term has come to mean anything from a hand-line fisherman, to a captain of a major fishing vessel, to the director of a small fleet, to the owners of a complex and integrated industry. At some level of integration, all actors on the scene can become significant members of a community that has an interest in the potential benefits of fishing. They represent social or economic units that have certain demands and reflect a variety of uses of the ecosystem. Not all of them depend on the landing of fish to argue that they have a significant interest or responsibility.

Society has, over the years, developed an effective model for dealing with such complex situations. Called by various names in different regions, it is basically a type of federated, multileveled system of administration, where the information flow within the system attempts to associate the array of demands with the array of apparent opportunities. Initially, both information and demands may emerge at the level of the community cultural or economic and social units in which human values receive basic expression. Under certain circumstances they are associated or merged in the larger and larger economic or political units through which human social activities are balanced and realized in the modern technological society.

Wilson and Dickie (1995) provide a more detailed discussion of the elements of structure that appear to be needed in fisheries. Here we wish only to draw attention to the fact, more and more generally recognized, that sustaining fisheries ecosystems requires, at the very base of our social and economic system of control in fisheries, a new, more effective and interactive means of sharing both the responsibility and the authority for regulation among the many levels of operation and management through which our major cultural values are expressed. There especially seems to be a new need for authority to be shared with associations of user groups at the most local operational and observational levels of the system.

Society faces many urgent ecological challenges as we enter the new millennium. Whether fisheries in particular, or the global oceanic ecosystems in general, will survive depends in large part on decisions made in the next few decades. Our survival as a species depends on our collective ability to make appropriate choices. It remains to be seen whether we can summon the will to construct or conserve our "sandpiles" intelligently, with some sense of what is at stake, or whether we will choose to tilt vacuously at windmills. The choice is ours, and the consequences are enormous.

## Summary

We have drawn attention to the practical difficulties of attempting to control a multileveled hierarchical system of interactions, both for identifying the causes of change and for forming remedial measures that could help sustain its production. The difficulties are particularly evident in exploited fisheries systems, where the economic and social forces that generate fishing effort issue from causes that are quite outside the negative feedback processes that maintain the balance within natural production systems. In such situations management cannot rely on feedback processes between fishing effort and catch to exert a restraining control on new fishing effort on the time scales at which biological production processes operate. In this chapter we have been concerned to clarify the implications of this lack of correspondence between the forces that motivate exploitation and the yields from natural fish production systems, in relation to traditional management principles and the organizations needed to implement them.

At the most general level of the production system, interactions between species and other stock elements may play a significant role in determining the resilience of the natural communities of organisms to exploitation. Traditional analysis conceives of production systems as summations of single-species models of production, applied at the syntactic level of effects. Such analysis is unable to offer information on production conceived as resulting from an interaction of population elements of various species in an energy-limited system—although the biomass spectrum suggests that the community on which fishing depends is of this type. Reanalysis of fisheries data on a spectral basis could offer information on this important practical question, but has not yet been undertaken.

The contrast between the reality of natural production system interactions and the current theory of their dynamics is most evident in the practical differences in approach resulting from the failure to distinguish between the syntactic and the semantic effects that act within systems.

Current models treat fishing mortality as a syntactic effect, resulting in a uniform rate of removal on the entire set of population elements, rather than as a semantic effect that exerts a local impact, tending to disrupt spawning concentrations, migration routes, and the patterns of predator-prey relations on which the organization of the energy flow depends. In fact, the new technologies of both fish finding and catching make local population elements—which in at least some cases sustain certain qualities of the stock—especially liable to disruption or even complete destruction. These effects are appreciated by fishermen, who observe the system at the local level. But they seem to be generally transparent to the usual modes of data collection and analysis on which management currently depends. At present it appears that effects at the semantic level have become so pervasive that the whole system output has been drastically affected.

We have discussed the serious management implications of the differences in perception between the principal participants in management—fishermen and administrators—regarding appropriate management. Because of the time delays between the fishermen's immediate perceptions of local deterioration of system organization and the administration's broad-scale measures of average population changes, differences between their points of view tend to become exaggerated, multiplying the distrust that inhibits effective control by centralized authority. Since both broad syntactic and local semantic effects may have separate but equally strong effects on energy flow in the system, we believe that the evidence for interactive energy systems offered by the biomass spectrum mandates a reordering of the entire management process in which authority and responsibility will be more evenly shared across the levels of system manipulation.

॰

# CHAPTER 11

# Final Observations

In this final chapter we review four topics in the interests of furthering the lines of inquiry opened in the book. First, we return to considering the extent to which findings for aquatic systems apply to terrestrial systems. In the two succeeding sections we discuss what we regard as relevant social and intellectual aspects of ecological analysis, resulting from the inevitable subjectivism that occurs in science during the choice and development of "appropriate" system models. Finally, we try to recognize places where we need to better understand and respond to the complexity that arises in ecosystem management and that must be taken into account to support its future development.

## Aquatic and Terrestrial System Studies

Pieces of the evidence for biomass spectral theory have appeared at various times during the past quarter of a century, in observations of aquatic and terrestrial ecosystems over much of the earthly globe. However, as mentioned in the preface, at the end of this review we find ourselves with a sense that, despite the breadth of the source material and of its apparent applicability to management, a comprehensive understanding of the role that the spectrum serves in ecology could scarcely have emerged much before now. The field in which current scientific thought is being developed represents a new intellectual movement toward a holism that seeks to comprehend hierarchical concepts of system behavior and theories of self-organization along with new theories of modeling. Together they show how it is possible to incorporate elements of observed actions on various levels of aggregation into a coherent causal network. This seems to be especially important to developments in the complex field of ecology, which is currently exploring new possible relationships between science and its usefulness to society.

The understanding of ecological processes we have assembled and assessed is in a far from perfect or complete state. Clearly it is liable to, and invites, further development. With vigilance and help from editors, referees, and friends, we have attempted to avoid outright errors or omissions of facts and interpretations regarding the biomass spectrum and its implications about the nature of aquatic ecosystems. However, it will be apparent to a majority of ecologists that because our coverage of terrestrial systems leaves so much unsaid, there must be room for significant additions and modifications. We treat this final chapter as an opportunity to point to some of the ways that the methodological synthesis in aquatic systems may be useful in establishing a broader understanding of terrestrial systems.

There can be little doubt that the two similarity principles—one involving predator-prey energy transformation and the other its transmission—that we have adopted as the basis for the dynamic theory of aquatic production must have equal importance for all organisms, whether in aquatic or terrestrial systems. With respect to the transformation processes, we have at several points drawn attention to the current confidence resulting from the discovery that architectural limits on energy transport within living organisms underlie the common relations of the metabolism of both plants and animals to their body size (Enquist et al. 1999). This accords with a recognition that it is the physiological limits of individual organisms that give rise to the integral aquatic biomass spectrum. These physiological limits must equally underlie energy utilization by individuals in terrestrial systems. Indeed, the evidence for common allometries of specific production over very diverse groups of organisms, originally pointed out by Banse and Mosher (1980), gives no indication of essential differences between aquatic and terrestrial systems. In light of the new evidence, the allometries must now be seen as common to all animals and plants.

The ecological significance of this understanding lies in the realization that what we have been calling the "physiological scaling" of the production system is, in fact, a reflection of the syntactic level of metabolic relationships that governs the generating mechanisms of all biological production. As pointed out by Rosen (1991), such a general set of relations arises because at this basic level of observation we are able to treat organisms as particles of living matter that are identical in terms of the relationships between their metabolism and their body size. That is, syntactic rules govern interactions of organisms in the manner of the laws of physics because this way of observing and analyzing the ecological system reduces the units of measurement to those of homogeneous particles.

A noteworthy consequence of this level of approach to analyzing aquatic systems is that the plot of individual organisms by body size in the integral biomass density spectrum is adequately described by the continuous flow model originally proposed by Platt and Denman (1977, 1978). Their ingenious model is an entirely appropriate basis for comparisons among whole systems and has been so used by a number of recent investigators. At this most reduced hierarchical level of observation, in keeping with the methods of the physical and mathematical sciences, comparative analysis gives rise to the basic allometric measures that are found among all assemblages of individual organisms. The allometric measures are the same relationships that have been studied as biological laws of metabolism for the century and a half since they were first recognized. Allometries at this level apply equally to organisms in aquatic and terrestrial systems.

By contrast, we have found that in order to understand causal relations in ecosystem dynamics it is essential to differentiate the problems of energy transformation *within* organisms from the problems of energy transmission *between* organisms, hence our concern with this aspect of predator-prey relations. This is the level of observation that we have termed an "ecological" scaling. From the discussion in foregoing chapters we can define aspects of this ecological scaling with Rosen as a semantic level of cause and effect. It is at this semantic level of predator-prey energy transmission that the oscillatory nature of the distribution of biomass density with body size appears in aquatic data. As we have pointed out, this level of interaction in energy flow results from the adjustments in density that are necessary if the nutritional needs of the predator density are to be harmonized with the production afforded by the prey density. That is, energy distribution in the ecosystem is a function of particular components made up of recognized groupings of the basic particles.

It is at this ecological, predator-prey, semantic level of the spectrum that the allometric relations between specific production and body size exhibit a slope different from that which characterizes the processes of energy transformation within individuals. Predator-prey density adjustments reflect effects additional to those defined by the metabolic limits to transmission within the bodies of individual organisms. We have pointed out, however, that evidence shows density adjustments to be common to aquatic and terrestrial organisms. The evidence of Banse and Mosher (1980) for the relatively steeper slopes of specific production with body size came from both terrestrial and aquatic species. In addition, evidence for adjustments in density in large, terrestrial grazing mammals was reflected in Damuth's (1981) detection of changes in the degree of the overlap of home range with body size. Clearly, the set of physiological and ecological factors that underlies much of the biomass spectrum must equally affect organisms in both aquatic and terrestrial environments. The mechanistic basis for energy transmission in the production functions of biological systems is unlikely to be different among different ecosystems.

We are left with the question of why there is such limited evidence for a spectrum of biomass domes in terrestrial systems. Holling (1992) thought it might be related to observer expectations. However, it is worth noting a particular aspect of our aquatic system theory that seemed to simplify model building. It was the hypothesis of a regularity in the predator-prey body-size ratio. In aquatic ecosystems this seems a reasonable hypothesis. The phenomenon of a gradation in body size throughout aquatic ecosystems is often considered an outcome of the importance of small phytoplankton as the productive base of the energy flow. In addition, the density of the aqueous medium gives rise to a three-dimensional distribution of organisms that seems to be quite different from that in the relatively two-dimensional terrestrial world. Thus, while density layers of the water in both freshwater and marine systems create interfaces that affect the distribution of both chemical and biological properties, the pelagic zone has a fluidity that imposes relatively little obligatory structural restriction on the distributions of organisms in it. By contrast, the character of land areas is strongly dependent on vegetational structures. These frequently gives rise to feeding territories, which are physical limits that offer a kind of insurance of energy availability in relation to the predator's survival and reproductive requirements. Thus while territory sizes, like metabolic requirements, are related to body size, such structural phenomena are likely to obscure aspects of the similarity principles that we believe make the feeding body-size ratio such an important part of the aquatic system (Ritchie and Olff 1999).

We conclude that, though the parameters of energy distribution and supply that underlie predator-prey transmission in aquatic and terrestrial systems may not be essentially different as syntactic effects, there seem to be semantic differences that arise from the physical structuring of aquatic and terrestrial systems. These must create special problems in sampling and classifying the energy transmission components in terrestrial systems. They may even obscure or obviate the perception of energy flow as a single, simple process describable by a biomass density spectrum in relation to body size. With growing confidence in the commonality of the underlying principles, however, it is reasonable to suppose that simple alternative methods will be found that parallel those of the aquatic spectra.

An important implication of the foregoing, with respect to applying the results of the hierarchy of allometric relations to management, is that analyses of either aquatic or terrestrial ecosystems cannot be adequately described at the syntactic level of observation alone. As Rosen (1991) might have put it, the concerns of managing biological systems are not sufficiently comprehended by the reduction of biological elements to their simplest common characteristics. It is the *components* of the natural systems, formed of significant patterns of groups of organisms, in either their spawning, migratory, or feeding behavior aggregations, that dominate their interactions with human exploiters. These same factors have a critical role in determining an effective management in both aquatic and terrestrial ecosystems.

In aquatic systems, fishermen have always understood that successful catching depended on intimate knowledge of the precise times and places for migrations or spawning aggregations. However, the scientific syntactic view of model building and administration failed to appreciate the significance of this level of knowledge. The importance of semantic knowledge may not have been so apparent in the cases of the hunters who originally swept down on the herds of bison or caribou that once seasonally dominated the landscapes of North America, or on the grazing herds on the grasslands in Africa. But in retrospect, much the same considerations must be applied to both terrestrial and aquatic environments. Their commonality becomes clearer once the wanton destruction of the large herds or of the massive fish schools has decimated initial stocks, threatening the survival of the human societies that depended on them, and demonstrating that greater knowledge and greater care were needed for continued exploitation. In both the aquatic and terrestrial systems, we have by now sufficiently witnessed the devastating power of human technology and tasted the weakness of our ability to take informed, hence responsible, action in time.

While these comparisons are simplistic, they make clear that our perceptions of our abilities to understand and control human activity are seriously out of step with our capacities to take temporary advantage of the opportunities available to us. In this sense it is somewhat ironic that Rosen should have adopted the term "semantic" to describe the level of observation necessary to close the gap between our ability to destroy and our still poorly tested ability to sustain natural production systems. However, the directing of our attention by the word "semantic" to the implicit concepts of "meaning" in interpreting the significance of the semantic, distributional-type information depends on our realizing that we need to learn how to observe systems at this more intimate level of interactions among component groups in ecosystems.

This semantic aspect of the hierarchical approach to analysis and management of natural production systems seems to be leading to new experiments in social participation and responsibility in the form of various "ecology action" groups. Previously, some of these groups have shown a militancy that reminds one of the frustration so often symptomatic of incipient and needed social change in the past. Perhaps it is a testimonial to the enhanced levels of education and standards of living that, with some local exceptions, there have already been a number of serious attempts to incorporate divergent points of view into a common social framework, now leading in the direction of supporting sustainable economic production systems. An increased knowledge of the general nature of systems may help us to understand the legitimacy of this new move toward responsible participation and how to take advantage of it before the frustrations of the "governed" have been expressed in violence.

However, the very nature of ecological studies restricts unequivocal testing of some of the ideas of system behavior by the kind of traditional scientific logic expressed in controlled laboratory experiments. In keeping with the geological and cosmological sciences, ecology is dependent on observation, comparison, and deduction; so the implications of any new set of observations may not be readily apparent or "objectively" testable in relation to theory. This has arisen in our text as we repeatedly resorted to complicating, semantic factors as aspects of explanation. The methodologies of violence may themselves have yielded new insights, by inadvertently intensifying observation of situations that had been neglected or insufficiently studied earlier. However, a more analytical, even contemplative and considered, approach to ecological analysis and management seems relatively easy to justify, and is more in keeping with current developments in the nature of scientific analysis. In what follows, we draw attention to a number of facets of this introspection that we believe have a role in developing biomass spectrum theory and justifying its use.

## Relations Between Science and Society

Oppenheimer (1958) pointed out that "Science starts with preconception, with the common culture, and with common sense. It moves on to observation, is marked by the discovery of paradox, and then is concerned with the correction of preconception." But, as he goes on to point out, where the results of scientific study are accepted back into society, they require for their understanding a kind of resonance from the community into which they enter. The flow of influences thus clearly moves in both directions; science is not an activity isolated from the common perceptions and aspirations of society. Science gets its initial ideas from what Oppenheimer calls the "preconceptions" of society, and, working in a particular way, may be enabled to confirm or correct them, provided that the requisite resonance develops in the culture. This is most popularly recognized in the history of the cosmology of Galileo and of Newton, both in the initial discoveries and in the processes of acceptance, which sometimes involved considerable difficulties for the first expositors. At the present critical state of the aquatic, environmental sciences, we need to be concerned with the nature of this exchange of influences.

We can clarify our interest in this process by referring to the nature and consequences of insights in twentieth-century scientific thought. Among the most prominent insights is the principle of complementarity, first elaborated by Niels Bohr in the 1920s in connection with the problem of understanding the mechanically irreconcilable wave and particle explanations of the phenomenon of light in physics. Its application to biology was thoughtfully discussed by him in an address to the International Congress on Light Therapy, reprinted in *Nature* (Bohr 1933). As Bohr pointed out, juxtaposition of the seeming contradictions of two logically inconsistent explanatory processes had an important impact on understanding the relation between the observer and the observed in science that still helps us appreciate the breadth of vision behind all of science and its application.

Oppenheimer (1958) pointed out that Bohr's concept of complementarity actually developed much earlier in connection with his boyhood interests in "the complementary character of the introspective and the behavioural description of man, in the complementary character of dealing with experience in the light of love and in the light of justice, and from the familiar yet disturbing tensions of comprehending in one description causal explanation of behavior and moral condemnation of behavior." In this succinct way, Oppenheimer makes alive for us something of the remarkable nature of Bohr's perception of human thought, which in this imagery is seen in relation to the field of influences in which it arises. That the same basic information finds such necessarily different treatment, and gives such different results, when viewed from the point of view of love or of justice, shows the potential for a new scientific point of view. It is in sharp contrast to the thinking of the nineteenth century in which isolated logical thought was largely regarded as the basis of objectivity. Beginning with the work of Bohr and the corresponding insights of Heisenberg's uncertainty principle in the 1920s, it can now be seen that the process of scientific discovery and understanding cannot be disconnected from the purposes for which observation, analysis, and thought are being marshaled. Particularly in relation to the life sciences, as Bohr (1933) himself puts it, "the concept of purpose, which is foreign to mechanical analysis, finds a certain field of application in problems where regard must be taken of the nature of life. In this respect, the role which teleological arguments play in biology reminds one of the endeavours, formulated in the correspondence argument, to take the quantum of action into account in a rational manner in atomic physics." Ulanowicz (1997), in an earlier book in this series, provides further arguments in support of the concept of *teleos* in living systems.

Bohr's sensitivity to the limitations of language led him to emphasize that his concepts of the difficulty of biological research have nothing to do with a return to notions of vitalism, which obscure rather than explain the phenomenon of life. What he makes equally clear, in keeping with points we have made earlier in this text, is that the alternative reduction of the description of physiological mechanisms of life to "clockworks" is not an explanation either. It is at best an analogy suitable only for the very limited purposes of a particular explainer in a particular context. "Taking apart" an organism or ecosystem for purposes of analysis has very different consequences than the taking apart of a clock. And this recognition may, in fact, assist the growing appreciation of organization as a relational phenomenon, which finds expression in the forms generated in the biological population structures that we observe. F. E. J. Fry was one of the first to point out that relations between organization and form are aspects of analysis that must be taken into particular account in biological, scientific analysis (Kerr 1990). Our perception of the biomass spectrum is that it carries with it the clear traces of this holistic nature of ecological phenomena.

The implications of this new realization that physics is an aspect of human endeavor perhaps reached their main importance in ecological analysis in connection with the choice and elaboration of system models. In developing a model, one must first be clear about its objectives or purposes. The second step concerns heuristically characterizing the model's organization. But in both steps we invoke influences that are outside the boundaries of the model itself. That is, both steps are functions of the investigator or observer, rather than a function of the observed system alone. Thus, Bohr's insights have cleared the way for a deeper recognition of something of what takes place in the flow of influences that Oppenheimer saw between the science and its society. We cannot neglect the essential relations between the requirements of scientific model building and the qualities of the observer.

Obviously, the process of study cannot take place without a qualified observer who undertakes to generalize on the myriad features that are observable. However, even at the level of the present study of aquatic production processes, the devastating failures in fisheries management are stark testimony to the inadequacies of using oversimplified, intellectual models to understand natural production processes. Throughout this book we have repeatedly argued for broadening the basis on which models of aquatic ecological systems are composed. The context of modern science makes clear that the background, qualifications, and experience of the investigator inevitably create a subjective element. However, given appropriate channels of communication, the society in which the work has to be developed can help establish criteria of realism and completeness that form the basis for discussion and consensus in the process.

There is little doubt in our minds that past failures in ecological modeling and in its application to management have partly resulted from the isolation of the intellectual and technical aspects of ecological science and its administration from the social system to which the results are supposed to apply. We believe that the social milieu that can use the results of scientific analysis must be involved in the process of model-building as near to the start as possible. The distinction often made in the past between science and its application may have been quite false. It may even be that the truly interactive nature of science is beginning to emerge in present-day environmental activism, which needs only patience as mechanisms are developed that will moderate activism's confrontational aspects in favor of a broader comprehension.

## The Nature of Analysis

Our concern with defining an "appropriate" mode of analysis parallels and partly arises from the above discussion. Rosen's (1991) heuristic exposition of the differences in the bases for model building between biology and physics harks back to perceptions that had arisen in midcentury. A general summary of the associated problems was provided in a book edited by Koestler and Smythies (1969), called, appropriately enough, *Beyond Reductionism,* which describes early efforts to define the concept of "system" and the implications of its "organization." Rosen (1999) draws attention to the central role of Gödel's theorem, the consequences of which are that the postulates of a logical system cannot ultimately be proven from within the system itself. In a sense this reinforces the perceptions of the importance of context offered by Bohr. But in a very practical additional sense, it raises important questions about how one decides how large a system is and how much detail is needed to capture the requirements for realism.

We are necessarily faced with questions about the scope of the models that have heretofore dominated aquatic ecology. In the context of the biomass spectrum, there are clear contrasts between the homogeneous particle, continuous energy-flow models of Platt and Denman and the discontinuous, predator-prey elaboration proposed by Thiebaux and Dickie and treated in detail in this account. We have already commented on their actual complementary usefulness for studying particular questions. In another case, that of specific fisheries applications of ecosystem theory, there are contrasts between the elaborations of the single-species, dynamic-pool, numerical models introduced by Baranov, and the perceptions that were embodied in Schaefer's logistic model approach. What concerns us here is how to learn from the combinations of these contrasting views, rather than dissipating energies in asserting the ascendancy of one over the other. In the development of the debate over applications of the dynamic-pool and logistic fisheries models, for example, there was a move, of questionable merit, to resolve the differences. It resulted in an attempt to reformulate the logistic approach so that it became primarily an alternative method of data handling for the situations comprehended by the dynamic pool but with fewer data (Schaefer and Beverton 1963). Consequently, the unique nature of the logistic as a potential approach to a multispecies population measure was obscured, although not entirely lost, in view of the multispecies virtual population analyses that have begun to appear in recent years (Gislason and Helgason 1985). This temporary misdirection is ironic in view of a present-day near consensus that one of the greatest needs in fisheries is for a method of appreciating the seemingly unpredictable interactions taking place in the whole ecological community.

There have been other approaches to the understanding of ecosystems, and will be many others. In relation to biomass spectral analysis, many of those in existence can be seen as representing one or the other of the continuous- or discontinuous-flow lines. In the contrast between existing approaches we may have found examples of the elements of paradox expected by Oppenheimer to characterize the development of any line of investigation. In alternative models in ecology generally, or in its more specific fisheries applications, we have the basic material to make a first step toward resolving apparent contradictions in the way we explain phenomena. As is implied by Bohr's insights, there is no requirement that the paradox between alternative approaches be resolved in the interests of espousing one view or another. What is essential is to differentiate the principles underlying the various perceptions so as to understand the basis for their capacity to explain and predict events at various levels of observation. The purpose of investigation is not so much to test the models as to understand their present utility and reliability as guides to management. The two basic approaches to the spectrum that have been defined to date emphasize different aspects of the dynamics of ecosystems, and these have different utilities for different situations. Understanding is made the richer by the presence of alternatives, if they are treated objectively. The example of complementarity suggests that objectivity in science does not rest as much with the creation of ideas of systems as with the dispassionate assessment of their worth in relation to the perceived purposes of the system in which they need to be applied.

What is of greater urgency for future analyses is to maintain an alert contrast between, on the one hand, the necessary practice of formulating and reformulating models, to capture salient dynamical features that will improve the reliability of extrapolation, and on the other hand, what has often happened in the past, which we can characterize only as a kind of tinkering with generalized concepts in an attempt to smooth over the differences between models in the interests of utilizing the suites of available data. More often than not this teasing or "tuning" of models to fit the data will be found to be an excuse for supporting entrenched interests by elaborating conventional data collections according to existing concepts. When a particular concept has not "worked," it is easier for the managers of systems to believe that the problem is "statistical," that more of the same will set things straight, than to face the problem of publicly admitting and explaining how the concept might have been deficient or wrong and trying to correct it. The self-protective motivation has played all too great a role in the management debacles of the great sea fisheries throughout the world. It needs to be recognized for what it is.

The identification of a commonly encountered aspect of this problem, the phenomenon of the "disappearing correlation," has been variously attributed to both David Cushing and John Gulland, once associates at the Lowestoft Fisheries Laboratory. They, or someone associated with them, drew attention to the annoying (and to some amusing) problem of the frequent disappearance of newly discovered correlations between environmental events and biological phenomena when more data are added. It was their perception, and the frequent experience of investigators, that this may often be the simple result of tinkering with data, without due regard to the scale on which the underlying phenomena are operating. However, the situation is considerably more serious than appears on the surface. Problems of this sort played a significant role in confusing the differentiation of fisheries and environmental effects in the Pacific Halibut controversy mentioned in chapter 8. It was not until Skud's (1975) analysis of phenomena underlying catch variations, carried out on the scales at which the distribution of abundance is established, that the futility of previous arguments could be seen. A similar phenomenon was encountered in various forms during the International Biological Program and in subsequent attempts to use the data (cf. Cyr and Peters 1996), and may often plague unwary researchers studying the effects of environment on biological dynamics where the actual mechanisms of interaction are unknown or are very complex (Myers 1998). The results of such studies undoubtedly show how part of the utility of any model is an appreciation of both the time and spatial scales over which a relationship can be expected to operate. This phenomenon is one important aspect of the problem exposed by the thinking of Rosen (1991, 1999), and one that we have utilized here in the form of the distinction between syntactic and semantic modes of action in systems. Much of the apparent progress in analyzing complex situations has been a result of modern methods of observation, which allow observations to be made often enough and at fine enough scales to permit interpretation of what was, only a generation ago, considered "statistical noise" in system response.

The realism of any representation of a complex system may appear as much in the appropriate scaling of the processes identified in model construction as in the examination of the results of different model versions. However, the reality of the configuration of models used in an analysis is insufficient to guarantee their utility. The difficulties of the additional problem of ensuring utility underline the importance of another phenomenon: namely, that science has for many years accepted that editors and referees must be satisfied with the value of potential contributions to the scientific literature. This recognized recourse to peers is accepted as a necessary step in assuring the intellectual authenticity of a line of logical analysis. However, publication of ideas does not guarantee either their acceptance or their applicability. This process depends on a larger process of testing and understanding in the larger community, what Oppenheimer called a "resonance" in that community.

What concerns us here is that applying ecological knowledge to questions of resource management that arise in particular social contexts poses a difficult problem of application that urgently needs to be addressed more effectively. The real questions of applying scientific analyses, which we only touched on in chapter 10, justify a careful reexamination of the elements and workings of the administrative organization of those results. The development of management initiatives based on scientific analyses may require radical, alternative approaches to arrive at human solutions. Organization at the level of the administration of management is as important to its effectiveness as is organization within the biological system to its productivity. The self-protective impulses to which social organization has shown itself to be so prone stifle the capacity of the system to learn from and adapt to new perceptions.

## Hierarchy as an Aspect of Dealing with Complexity

In our work on the biomass density spectrum we have emphasized that our conclusions depend on a conception of the whole of a given aquatic ecosystem as having distinctive properties at at least three levels of aggregation of data. In our analyses these reflected different aspects of system variation in relation to body size. Such analysis enables an appreciation of a number of the factors underlying the dynamics of aquatic ecosystems, and that may exist in terrestrial systems as well. We identified these labels as (1) individual organisms, (2) trophic functions, and (3) what might be called species-type interactions. The reality of phenomena at these levels as parts of the operation of the whole system seems to be well established from the particular calculations made in different aquatic ecosystems. But as a generality this hierarchical method of analysis of an accepted whole does not conform to the traditional methods, in which the explanation of the appearance of general phenomena is sought at lower and lower levels of organization, ultimately at the level of the laws of physics and chemistry. The idea behind our analyses of a constraint on the parts by the whole may not be new, but is at least distinctive enough to have justified an extensive and penetrating discussion of the topic in a recent book by Ahl and Allen (1996).

In his early works, Rosen (1978) drew particular attention to the importance of classification in the analysis of biological systems. Classification, because it is imposed by the observer on the system, introduces the subjectivity we have already referred to. However, as Ahl and Allen point out, the results of classification in hierarchies show that the concept of whole systems allows comparisons that reveal how systems vary in practically important characteristics. There are differences in the relation of parts to the whole. In particular, in some systems the "whole" does not constrain certain of its parts. In such cases, predicting the actions of the whole from such parts cannot be a simple matter, nor can conclusions be based on arbitrary sets of observations.

One needs to recognize that in the complex field of ecology a degree of indeterminancy may recur. At present, for example, we must appreciate, as stated in chapter 8, that knowing the production in an area of the sea cannot yet tell us what species will be involved, any more than the presence of an arbitrary set of species can ensure a given level of production. Still, species possess characteristics that enable them to fit a particular environmental pattern, and to function in relation to others that are present. It may well be that stability in ecosystems is a function of redundancy of species, such as appears often to be the case in tropical systems. Perhaps multispecies systems will always respond to exploitative challenges better than those we are inclined to refer to as depauperate, northern systems. The key point is that no one conceptual pattern is likely to capture all possible variation in observations. Thus, for example, an energy-based model may yield quite different conclusions about the importance of individual species than an information-based model of the same system. Analysis intended for application to human problems, in particular, needs a level of ingenuity at least equal to that of the initial perceivers of the problems.

Through all the arguments about the complexity found within whole systems of organisms, generalities can still be made about the suitability of various species to particular environments. The continued success of some species in colonizing new environments demonstrates that the present ecological situation of the world is not uniquely ordained, nor is it necessarily the ultimately suitable configuration possible. For example, where biota have been transported in new ways— such as by the unwitting flushing of ballast tanks of ships, which are filled in European or Asiatic waters and emptied in the Great Lakes, or deliberately by the introduction of Pacific Salmon into the same lakes—

the sudden and unexpected flourishing of new species is testimony to their continuing adaptability to new environmental conditions. Nevertheless, we are rarely in a position, even in these relatively simple cases, to identify in advance the salient features of a given species or environment that will make the difference between success or failure in a given "introduction." This uncertainty makes it more understandable that we have so far been unable to predict the species changes accompanying the vast increases in rate of exploitation in local fisheries during the past quarter century. We still cannot tell whether prediction will be possible with more detail, or whether the perception of real distinctions will emerge only from alternative classifications or typologies of systems.

We believe two very different points of view will continue to dominate discussions about this and other, parallel cases of change and potential change in both the environment and the fauna associated with it: is stasis the desired norm, or is change? They are an example of a fundamental dichotomy that arises in any so-called worldview. We must recognize this inevitable tendency, because recognition is surely the first step toward Oppenheimer's "resonance." Following Bohr, we need to anticipate that bridging the gap between the differences that inevitably arise at any given level of understanding can be the equivalent of recognizing and bridging the dichotomy between love and justice. Appreciating the depth of such differences may be an essential element in defusing or resolving disputes before we are faced with the more difficult problem of dealing with the consequences of violence.

An equally important approach to appreciating the possibly fundamental differences in judging phenomena is that taken by Bohm (1980). He drew attention to the shifts in emphasis, through time, in such fundamental concepts as measurement. In his view, among the ancient Greeks the notion of measure went "far beyond that of comparison with an external standard, to point to a universal sort of inner ratio of proportion, perceived both through the senses and through the mind." With time, however, this Greek emphasis on measure as a form of insight was replaced in the modern Western world by a greater concern with practical manifestations of things and events in relation to some arbitrary, external criterion. As he put it, measurement eventually appeared as "absolute truths about reality as it is, which men seemed always to have known, and whose origin was often explained mythologically as binding injunctions of the Gods. . . . [T]he forms induced in perception by this thought were now seen as directly observed objective realities, which were essentially independent of how they were thought about" (20–22).

These transformations of ideas are assuming special, practical importance in a world where it is widely perceived that the increasing density of human beings is upsetting a former balance in the established living order. Bohm's conclusions about the need to recognize the presence of an internalized, "implicate order," which can explain and resolve the apparently strange dichotomies and contradictions in the ordinary level of perceptions, are clearly outside the domain of science as it was conceived by nineteenth-century thinkers. Nevertheless his probing of the depths of attitude for the sources of what often seem simple, ordinary, but contradictory concepts is by no means irrelevant to the problems that arise in the application of science to human affairs. Such an adventurous branching out of scientific thought, in fact, helps explain the subtlety in the processes of perception that underlie the different opinions about the world.

In areas of human endeavor motivated by processes as emotional and sensitive as the concept "environment," the origin of attitudes and their impact on the creation of system models and on the testing of the models in relation to management objectives is no longer a philosophical frill outside the purview of science. Even pragmatic acceptance in our society of fundamentally different approaches to environment, such as are now frequently appearing between "managers" and "the public," requires a patience and breadth of viewpoint that challenges the foundations of the political and administrative institutions charged with ensuring orderly development and sharing of resources. We are forced to take radically new viewpoints into account if we expect the support of an increasingly informed and highly intelligent public. When this public senses the loss of a broader purpose, it participates in increasingly bizarre, reactive, unconscious, sometimes chaotic actions, despite the fact that both managers and public believe that they are engaged in protecting the long-term human use of natural resources! How to recognize and acknowledge values that can bridge the inevitable gaps is a central problem for ecology, as a small branch of the larger human endeavor to understand the place of the parts in the whole.

Ahl, V. and T. F. H. Allen. 1996. *Hierarchy Theory: A Vision, Vocabulary, and Epistemology*. New York: Columbia University Press.

Ahrens, M. A. and R. H. Peters. 1991a. Patterns and limitations in limnoplankton size spectra. *Can. J. Fish. Aquat. Sci.* 48:1967–1978.

———. 1991b. Plankton community respiration: Relationships with size distributions and lake trophy. *Hydrobiologia* 224:77–87.

Allee, W. C. 1931. *Animal Aggregations: A Study in General Sociology*. Chicago: University of Chicago Press.

Allen, T. F. H. and T. B. Starr. 1982. *Hierarchy: Perspectives for Ecological Complexity*. Chicago: University of Chicago Press.

Amarasekare, P. 1998. Allee effects in metapopulation dynamics. *Amer. Naturalist* 152:298–302.

Andersen, K. P. and E. Ursin. 1978. A multispecies analysis of the effects of variations of effort upon stock composition of eleven North Sea fish species. *Rapp. Proc.-verb. Réun. Cons. Int. Explor. Mer* 172:286–291.

Apollonio, S. In preparation. *Searching for Systems in the Gulf of Maine*. New York: Columbia University Press.

Ashby, W. R. 1964. *An Introduction to Cybernetics*. London: Methuen.

Bagenal, T. B. 1977. Effects of fisheries on Eurasian perch (*Perca fluviatilis*) in Windermere. *J. Fish. Res. Board Canada* 34:1764–1768.

Bak, P., C. Tang, and K. Wisenfeld. 1988. Self-organized criticality. *Physical Review A.* 38:364–374.

Banse, K. and S. Mosher. 1980. Adult body mass and annual production/biomass relationships of field populations. *Ecol. Monogr.* 50:355–379.

Baranov, F. I. 1918. On the question of the biological basis of fisheries (in Russian). Nauchnyi Issledovatelskii Ikhtiologicheskii Institut. *Izvestiia* 1:81–128. [Translated by Natalie B. Notkin, 1934.]

Bax, N. J. 1991. A comparison of the fish biomass flow to fish, fisheries, and mammals in six marine ecosystems. *ICES Mar. Sci. Symp.* 193:217–224.

———. 1998. The significance and prediction of predation in marine fisheries. *ICES J. Mar. Sci.* 55:997–1030.

Beers, J. R., F. M. Reid, and G. L. Stewart. 1982. Seasonal abundance of the microplankton populations of the North Pacific Central Gyre. *Deep Sea Res.* 29:217–245.

Bentzen, P., C. T. Taggart, D. E. Ruzzante, and D. Cook. 1996. Microsatellite polymorphism and the population structure of cod (*Gadus morhua*) in the North West Atlantic. *Can. J. Fish. Aquat. Sci.* 53:2706–2721.

Beverton, R. J. H. 1990. Small marine pelagic fish and the threat of fishing; are they endangered? *J. Fish Biol.* 37(Suppl. A):5–16.

Beverton, R. J. H. and S. J. Holt. 1957. *On the Dynamics of Exploited Fish Populations.* U.K. Ministry for Agriculture and Fisheries, Fisheries Investigations, 2d ser., vol. 19.

Blaikie, H. B. and S. R. Kerr. 1996. Effect of activity level on apparent heat increment in Atlantic cod, *Gadus morhua. Can. J. Fish. Aquat. Sci.* 53:2093–2099.

Bohm, D. 1980. *Wholeness and the Implicate Order.* London: Routledge & Kegan Paul. Ark edition, London: Routledge, 1995.

Bohr, N. 1933. Light and life. *Nature* 19:421–423 (March 25) and 457–459 (April 1).

Bonner, J. T. 1965. *Size and Cycle, an Essay on the Structure of Biology.* Princeton, N.J.: Princeton University Press.

Booke, H. E. 1981. The conundrum of the stock concept—are nature and nurture definable in fishery science? *Can. J. Fish. Aquat. Sci.* 38:1479–1480.

Borgmann, U. 1983. Effect of somatic growth and reproduction on biomass transfer up pelagic food webs as calculated from particle-size-conversion efficiency. *Can. J. Fish. Aquat. Sci.* 40:2010–2018.

———. 1987. Models of the slope of, and biomass flow up, the biomass size spectrum. *Can. J. Fish. Aquat. Sci.* 44(Suppl. 2):136–140.

Boudreau, P. R. and L. M. Dickie. 1989. A biological model of fisheries production based on physiological and ecological scalings of body size. *Can. J. Fish. Aquat. Sci.* 46:614–623.

———. 1992. Biomass spectra of aquatic ecosystems in relation to fisheries yield. *Can. J. Fish. Aquat. Sci.* 49:1528–1538.

Boudreau, P. R., L. M. Dickie, and S. R. Kerr. 1991. Body-size spectra of production and biomass as system-level indicators of ecological dynamics. *J. Theor. Biol.* 152:329–339.

Bowen, W. D. 1997. Role of marine mammals in aquatic ecosystems. *Mar. Ecol. Progr. Series* 158:267–274.

Brander, K. and P. C. F. Hurley. 1992. Distributions of early-stage Atlantic cod (*Gadus morhua*), haddock (*Melanogrammus aeglefinus*), and witch flounder (*Glyptocephalus cynoglossus*) eggs on the Scotian Shelf: A reappraisal of evidence on the coupling of cod spawning and plankton production. *Can. J. Fish. Aquat. Sci.* 49:238–251.

Brandt, S. B., D. M. Mason, E. V. Patrick, R. L. Argyle, L. Wells, P. A. Unger, and D. J. Stewart. 1991. Acoustic measures of the abundance and size of pelagic planktivores in Lake Michigan. *Can. J. Fish. Aquat. Sci.* 48: 894–908.

Brett, J. R. 1965. The relation of size to rate of oxygen consumption and sustained swimming speed of sockeye salmon (*Oncorhynchus nerka*). *J. Fish. Res. Board Canada* 22:1491–1501.

Brodie, P. F. 1975. Cetacean energetics, an overview of intraspecific size variation. *Ecology* 56:152–161.

Brodie, W. B., S. J. Walsh, and D. B. Atkinson. 1998. The effect of stock abundance on range contraction of yellowtail flounder (*Pleuronectes ferruginea*) on the Grand Bank of Newfoundland in the Northwest Atlantic from 1975 to 1995. *J. Sea Res.* 39:139–152.

Brown, B. E., J. A. Brennan, M. D. Grosslein, E. G. Heyerdahl, and R. C. Hennemuth. 1976. The effect of fishing on the marine finfish biomass in the Northwest Atlantic from the Gulf of Maine to Cape Hatteras. *ICNAF Res. Bull.* 12:49–68.

Burkenroad, M. D. 1948. Fluctuations in abundance of Pacific halibut. *Bull. Bingham Oceanogr. Coll.* 11:81–129.

Calder, W. A., III. 1984. *Size, Function, and Life History.* Cambridge, Mass.: Harvard University Press.

Calkins, T. P. 1961. Measures of population density and concentration of fishing effort for yellowfin and skipjack tuna in the eastern Pacific Ocean, 1951–1959. *Bull. Inter-Amer. Trop. Tuna Comm.* 6:70–152.

Calow, P. 1977. Conversion efficiencies in heterotrophic organisms. *Biol. Rev.* 52: 385–409.

Campbell R. N. 1979. Ferox trout, *Salmo trutta* L., and charr, *Salvelinus alpinus* (L.), in Scottish lochs. *J. Fish Biol.* 14:1–29.

Carbone, C., G. M. Mace, S. C. Roberts, and D. W. MacDonald. 1999. Energetic constraints on the diet of terrestrial carnivores. *Nature* 402:286–288.

Chadwick, E. M. P. 1976. Ecological fish production in a small Precambrian shield lake. *Env. Biol. Fish.* 1:13–60.

Chapman, R. P. 1967. Sound scattering in the ocean. *Underwater Acoust.* 2: 161–183.

Claireaux, G., D. M. Webber, S. R. Kerr, and R. G. Boutilier. 1995. Physiology and behaviour of free swimming Atlantic cod, *Gadus morhua,* facing fluctuating salinity and oxygenation conditions. *J. Exp. Biol.* 198:61–69.

Clark, C. W. 1985. *Bioeconomic Modelling and Fisheries Management.* New York: Wiley.

———. 1990. *Mathematical Bioeconomics: The Optimal Management of Renewable Resources.* 2d ed. New York: Wiley.

Clark, J. R. 1952. Experiments on the escape of undersized haddock through otter trawls. *U.S. Fish and Wildlife Service, Commercial Fisheries Review* 14:1–7.

Clay, C. S. and H. Medwin. 1977. *Acoustical Oceanography: Principles and Applications.* New York: Wiley.

Cohen, E. B., M. D. Grosslein, M. P. Sissenwine, F. Steimle, and W. R. Wright. 1982. Energy budget of Georges Bank. In M. C. Mercer, ed., *Multispecies Approaches to Fisheries Management Advice,* pp. 95–107. Canadian Special Publication of Fisheries and Aquatic Science, no. 59. Ottawa.

Collie, J. S. and P. C. Spencer. 1994. Modeling predator-prey dynamics in a fluctuating environment. *Can. J. Fish. Aquat. Sci.* 51:2665–2672.

Colton, J. B. 1955. Spring and summer distribution of haddock on Georges Bank. U.S. Department of the Interior, Fish and Wildlife Service, Special Scientific Report, Fisheries No. 156.

Conover, R. J. 1978. Transformation of organic matter. In O. Kinne, ed., *Marine Ecology IV: Dynamics,* pp. 221–499. New York: Wiley.

Conover, R. J. and C. M. Lalli. 1974. Feeding and growth in *Clione limacina* (Phipps), a pteropod mollusc. II. Assimilation, metabolism, and growth efficiency. *J. Exp. Mar. Biol. Ecol.* 16:131–154.

Cook, R. M., A. Sinclair, and G. Stefánsson. 1997. Potential collapse of North Sea cod stocks. *Nature* 385:521–522.

Cottingham, K. L. 1999. Nutrients and zooplankton as multiple stressors of phytoplankton communities: Evidence from size structure. *Limnol. Oceanogr.* 44:810–827.

Craig, R. E. and S. T. Forbes. 1969. Design of a sonar for fish counting. *Fish. Dir. Ser. Havunders.* 15:210–219.

Creco, V. and W. J. Overholtz. 1990. Causes of density-dependent catchability for Georges Bank haddock *Melanogrammus aeglefinus. Can. J. Fish. Aquat. Sci.* 47:385–394.

Cyr, H. and M. L. Pace. 1993. Allometric theory: Extrapolations from individuals to communities. *Ecology* 74:1234–1245.

Cyr, H. and R. H. Peters. 1996. Biomass-size spectra and the prediction of fish biomass in lakes. *Can. J. Fish. Aquat. Sci.* 53:994–1006.

Daan, N. 1980. A review of replacement of depleted stocks by other species and the mechanisms underlying such replacement. *Rapp. Proc.-verb. Réun. Cons. Int. Explor. Mer* 177:405–421.

Damuth, J. 1981. Home range, home range overlap, and species energy use among herbivorous mammals. *Biol. J. Linnean Soc.* 15:185–193.

Damuth, J. D. 1998. Common rules for animals and plants. *Nature* 395:115–116.

De Aracama, J. D. 1992. Length-frequency distributions of demersal fishes on the Scotian Shelf. M.Sc. thesis, Biology Department, Dalhousie University, Halifax, N.S.

Denman, K. L., H. J. Freeland, and D. L. Mackas. 1989. Comparisons of time scales for biomass transfer up the marine food web and coastal transport processes. *Can. Spec. Publ. Fish. Aquat. Sci.* 108:255–264.

Dickie, L. M. 1972. Food chains and fish production. International Commission for the Northwest Atlantic Fisheries, Special Publication no. 8:201–221.

———. 1976. Predation, yield, and ecological efficiency in aquatic food chains. *J. Fish. Res. Board Canada* 33:313–316.

Dickie, L. M. and P. R. Boudreau. 1987. Comparison of acoustic reflections from spherical objects and fish using a dual-beam echosounder. *Can. J. Fish. Aquat. Sci.* 44:1915–1921.

Dickie, L. M., S. R. Kerr, and P. R. Boudreau. 1987. Size-dependent processes underlying regularities in ecosystem structure. *Ecolog. Monogr.* 57:233–250.

Dickie, L. M., S. R. Kerr, and P. Schwinghamer. 1987. An ecological approach to fisheries assessment. *Can. J. Fish. Aquat. Sci.* 44(Supp. II):67–74.

Doubleday, W. G. and D. Rivard, eds. 1981. *Bottom Trawl Surveys.* Canadian Special Publication of Fisheries and Aquatic Sciences, no. 58. Ottawa.

Duplisea, D. E. 1998. Structuring of benthic communities, with a focus on size-spectra. Doctoral diss., Department of Systems Ecology, Stockholm University, Stockholm, Sweden.

Duplisea, D. E. and S. R. Kerr. 1995. Application of a biomass size spectrum model to demersal fish data from the Scotian Shelf. *J. Theor. Biol.* 177:263–269.

Duplisea, D. E., S. R. Kerr, and L. M. Dickie. 1997. Demersal fish biomass size spectra on the Scotian Shelf, Canada: Species replacement at the shelfwide scale. *Can. J. Fish. Aquat. Sci.* 54:1725–1735.

Edmonds, A. 1974. *Voyage to the Edge of the World.* Toronto: McClelland and Stewart.

Engelmann, M. D. 1966. Energetics, terrestrial field studies, and animal produc-
    tivity. *Adv. Ecol. Res.* 3:73–115.
Enquist, B. J., J. H. Brown, and G. B. West. 1998. Allometric scaling of plant
    energetics and population density. *Nature* 395:163–165.
Enquist, B. J., G. B. West, E. L. Charnov, and J. H. Brown. 1999. Allometric
    scaling of production and life-history variation in vascular plants. *Nature*
    401:907–911.
Evans, D. O. and D. H. Loftus. 1987. Colonization of inland lakes in the Great
    Lakes region by rainbow smelt, *Osmerus mordax:* Their freshwater niche
    and effects on indigenous species. *Can. J. Fish. Aquat. Sci.* 44(Suppl. 2):
    249–266.
Fenchel, T. 1974. Intrinsic rate of natural increase: The relationship with body
    size. *Oecologia* 14:317–326.
Fogarty, M. J. 1995. Chaos, complexity and community management of fisheries:
    An appraisal. *Marine Policy* 19:437–444.
Fogarty, M. J., E. B. Cohen, W. L. Michaels, and W. W. Morse. 1991. Predation
    and the regulation of sand lance populations: An exploratory analysis.
    *ICES Mar. Sci. Symp.* 193:120–124.
Foote, K. G., A. Aglen, and O. Nakken. 1986. Measurement of fish target
    strength with a split-beam echo sounder. *J. Acoust. Soc. Am.* 80(2):612–
    621.
Fromentin, J. C., N. C. Stenseth, J. Gjøsæter, O. N. Bjørnstad, W. Falck, and
    T. Johannessen. 1997. Spatial patterns of the temporal dynamics of three
    gadoid species along the Norwegian Skagerrak coast. *Mar. Ecol. Progr.
    Series* 155:209–222.
Fromentin, J. C., N. C. Stenseth, J. Gjøsæter, T. Johannessen, and B. Planque.
    1998. Long-term fluctuations in cod and pollack along the Norwegian
    Skagerrak coast. *Mar. Ecol. Progr. Series* 162:265–278.
Frost, B. W. and L. E. McCrone. 1974. Vertical distribution of zooplankton and
    myctophid fish at Canadian weather station P, with description of a new
    multiple net trawl. *Proc. Int. Conf. Engineer. Oceanogr. Environ.* (Halifax,
    N.S.) 1:159–165.
Frost, T. M., S. R. Carpenter, A. R. Ives, and T. K. Kratz. 1995. Species compen-
    sation and complementarity in ecosystem function. In J. H. Lawton and
    C. G. Jones, *Linking Species and Ecosystems,* pp. 224–239. New York:
    Chapman and Hall.
Fry, F. E. J. 1947. Effects of the environment on animal activity. University of
    Toronto Studies, Biological Series no. 55. Publications of the Ontario
    Fisheries Research Laboratory, no. 68, 1–62.
———. 1949. Statistics of a lake trout fishery. *Biometrics* 5:27–67.

————. 1957. The aquatic respiration of fishes. In M. E. Brown, ed., *The Physiology of Fishes,* vol. 1, pp. 1–63. New York: Academic Press.

————. 1971. The effect of environmental factors on the physiology of fish. In W. S. Hoar and D. J. Randall, eds., *Fish Physiology,* vol. VI, pp. 1–98. New York: Academic Press.

Fuhrman, J. A. 1999. Marine viruses and their biogeochemical and ecological effects. *Nature* 399:541–548.

Gaedke, U. 1993. Ecosystem analysis based on biomass size distributions: A case study of a plankton community in a large lake. *Limnol. Oceanogr.* 38: 112–127.

Gasol, J. M., R. Guerrero, and C. Pedrós-Alió. 1991. Seasonal variations in size structure and procaryotic dominance in sulferous Lake Cisó. *Limnol. Oceanogr.* 36:860–872.

Gerlach, S. A., A. E. Hahn, and M. Schrage. 1985. Size spectra of benthic biomass and metabolism. *Mar. Ecol. Progr. Series* 26:161–173.

Gin, K. Y. H., J. Guo, and H.-F. Cheong. 1998. A size-based ecosystem model for pelagic waters. *Ecol. Modelling* 112:53–72.

Gislason, H. and Helgason, T. 1985. Species interactions in assessment of fish stocks with special application to the North Sea. *Dana* 5:1–44.

Gordon, H. S. 1954. The economic theory of a common property resource. *J. Agric. Econ.* 22:124–142.

Graham, H. W. 1952. Mesh regulation to increase the yield of the Georges Bank haddock fishery. International Commission for the Northwest Atlantic Fisheries, 2d Ann. Rep., part 3, 23–33.

————. 1954. United States Research in the Convention Area during 1953. *Int. Comm. N.W. Atl. Fish. Annual Proc.* 4:56–58.

————. 1957. United States Research, 1956. *Int. Comm. N.W. Atl. Fish. Annual Proc.* 7:63–64.

Griffiths, D. 1998. Sampling effort, regression method, and the shape and slope of size-abundance relations. *J. Animal Ecol.* 67:795–804.

Gulland, J. A. 1955. Estimation of growth and mortality in commercial fish populations. *Fish. Invest.* (London), Series II, 18(9):1–46.

————. 1962. The application of mathematical models of fish populations. In E. D. LeCren and M. W. Holdgate, eds., *The Exploitation of Natural Animal Populations,* pp. 204–220. Oxford: Blackwell.

————. 1983. *Fish Stock Assessment: A Manual of Basic Methods.* New York: Wiley.

Hanson, J. M., E. E. Prepas, and W. C. Mackay. 1989. Size distribution of the macroinvertebrate community in a freshwater lake. *Can. J. Fish. Aquat. Sci.* 46:1510–1519.

Hardy, A. C. 1924. The herring in relation to its animate environment. Part I. The food and feeding habits of the herring. *Fish. Invest.* (London), Series II, 7(3):1–53.

Harestad, A. S. and F. L. Bunnell. 1979. Home range and body weight— a reevaluation. *Ecology* 60:389–402.

Hargrave, B. T., G. C. Harding, K. F. Drinkwater, T. C. Lambert, and W. G. Harrison. 1985. Dynamics of the pelagic food web in St. Georges Bay, southern Gulf of St. Lawrence. *Mar. Ecol. Progr. Series* 20:221–240.

Harrison, S. and A. Hastings. 1996. Genetic and evolutionary consequences of metapopulation structure. *Trends Ecol. Evol.* 11:180–183.

Hastings, A. and S. Harrison. 1994. Metapopulation dynamics and genetics. *Ann. Rev. Ecol. Syst.* 25:167–188.

Heatwole, H. and R. Levins. 1973. Biogeography of the Puerto Rican Bank: Species turnover on a small cay, Cayo Ahogado. *Ecology* 54:1042–1055.

Hemmingsen, A. M. 1960. Energy metabolism as related to body-size and respiratory surfaces, and its evolution. *Reports of the Steno Memorial Hospital and the Nordisk Insulinlaboratorium* 9(Part II):7–110.

Hennemuth, R. C. 1979. Man as a predator. In G. P. Patil and M. L. Rosenzweig, eds., *Contemporary Quantitative Ecology and Related Econometrics,* pp. 507–532. Fairland, Md.: International Cooperative Publishing House.

Herman, A. W. 1992. Design and calibration of a new optical plankton counter capable of sizing small zooplankton. *Deep Sea Res.* 39:395–415.

Hewson, L. C. 1959. A seven-year study of the fishery for lake whitefish, *Coregonus clupeaformis,* on Lake Winnipeg. *J. Fish. Res. Board Canada* 16:107–120.

————. 1960. A history of the Lake Winnipeg fishery for whitefish, *Coregonus clupeaformis,* with some references to its economics. *J. Fish. Res. Board Canada* 17:625–639.

Hilborn, R. and C. J. Walters. 1992. *Quantitative Fisheries Stock Assessment: Choice, Dynamics and Uncertainty.* New York: Chapman and Hall.

Hochachka, P. W. and G. N. Somero. 1971. Biochemical adaptation to the environment. In W. S. Hoar and D. J. Randall, eds., *Fish Physiology,* vol. VI, pp. 99–156. New York: Academic Press.

Holčík, J. 1977. Changes in fish community of Klíčava Reservoir with particular reference to Eurasian perch (*Perca fluviatilis*), 1957–72. *J. Fish. Res. Board Canada* 34:1734–1747.

Holden, M. J. 1978. Long-term changes in the landings of fish from the North Sea. *Rapp. Proc.-verb. Réun. Int. Explor. Mer* 172:11–26.

Holling, C. S. 1986. The resilience of terrestrial ecosystems: Local surprise and global change. In W. C. Clark and R. E. Munn, eds., *Sustainable Develop-*

*ment of the Biosphere,* pp. 292–317. Cambridge: Cambridge University Press.

———. 1992. Cross-scale morphology, geometry, and dynamics of ecosystems. *Ecolog. Monogr.* 62(4):447–502.

Humphreys, W. F. 1979. Production and respiration in animal populations. *J. Animal Ecol.* 48:427–453.

———. 1981. Towards a simple index based on live-weight and biomass to predict assimilation in animal populations. *J. Animal Ecol.* 50:543–561.

Huntley, M. E., M. Zhou, and W. Nordhausen. 1995. Mesoscale distribution of zooplankton in the California Current in late spring, observed by Optical Plankton Counter. *J. Mar. Res.* 53:647–674.

Huntsman, A. G. 1918. Histories of new food fishes. I. The Canadian plaice. *Bull. Biol. Board Canada* 1:1–32.

———. 1948. Method in ecology—biapocrisis. *Ecology* 29:30–42.

———. 1962. Method in ecology—ectology. *Ecology* 43:552–556.

Hutchings, J. A. 1996. Spatial and temporal variation in the density of northern cod and a review of hypotheses for the stock's collapse. *Can. J. Fish. Aquat. Sci.* 53:943–962.

Hutchings, J. A. and R. A. Myers. 1994. What can be learned from the collapse of a renewable resource? Atlantic cod, *Gadus morhua,* of Newfoundland and Labrador. *Can. J. Fish. Aquat. Sci.* 51:2126–2146.

Hutchinson, G. E. 1953. The concept of pattern in ecology. *Proc. Acad. Nat. Sci. Philadel.* 105:1–12.

———. 1957. Concluding remarks. *Cold Spring Harbor Symp. Quant. Biol.* 22:415–427.

Hutchinson, G. E. and R. H. MacArthur. 1959. A theoretical ecological model of size-distribution among species of animals. *Amer. Naturalist* 93:117–125.

Huxley, J. S. [1932] 1972. *Problems of Relative Growth.* Reprint, New York: Dover.

Ivlev, V. S. 1960. On the utilization of food by planktophage fishes. *Bull. Math. Biol.* 22:371–389.

———. 1961. *Experimental Ecology of the Feeding of Fishes.* Translated by Douglas Scott. New Haven, Conn.: Yale University Press.

Johnson, L. 1972. Keller Lake: Characteristics of a culturally unstressed salmonid community. *J. Fish. Res. Board Canada* 29:731–740.

———. 1994. Long-term experiments on the stability of two fish populations in previously unexploited arctic lakes. *Can. J. Fish. Aquat. Sci.* 51:209–225.

Jones, R. 1973. Density dependent regulation of the numbers of cod and haddock. *Rapp. Proc.-verb. Réun. Cons. Int. Explor. Mer* 164:156–173.

———. 1978. Competition and co-existence with particular reference to gadoid fish species. *Rapp. Proc.-verb Réun. Cons. Int. Explor. Mer* 172: 292–300.

Kay, J. J., L. A. Graham, and R. E. Ulanowicz. 1989. A detailed guide to network analysis. In F. Wulff, J. G. Field, and K. H. Mann, eds., *Network Analysis in Marine Ecology: Methods and Applications,* pp. 15–61. Berlin: Springer-Verlag.

Keast, A. 1985. Planktivory in a littoral-dwelling lake fish association: Prey selection and seasonality. *Can. J. Fish. Aquat. Sci.* 42:1114–1126.

Kerr, S. R. 1971a. Analysis of laboratory experiments on growth efficiency of fishes. *J. Fish. Res. Board Canada* 28:801–808.

———. 1971b. Prediction of growth efficiency in nature. *J. Fish. Res. Board Canada* 28:809–814.

———. 1971c. A simulation model of lake trout growth. *J. Fish. Res. Board Canada* 28:815–819.

———. 1974. Theory of size distribution in ecological communities. *J. Fish. Res. Board Canada* 31:1859–1862.

———. 1979. Prey availability, metaphoetesis, and the size structures of lake trout stocks. *Investigacions Pesqueras* 43:187–198.

———. 1980. Niche theory in fisheries ecology. *Trans. Amer. Fish. Soc.* 109: 254–260.

———. 1982. Estimating the energy budgets of actively predatory fishes. *Can. J. Fish. Aquat. Sci.* 39:371–379.

———. 1990. The Fry paradigm: Its significance for contemporary ecology. *Trans. Amer. Fish. Soc.* 119:779–785.

Kerr, S. R. and N. V. Martin. 1970. Trophic-dynamics of lake trout production systems. In J. H. Steele, ed., *Marine Food Chains,* pp. 365–376. Berkeley: University of California Press.

Kerr, S. R. and R. A. Ryder. 1977. Niche theory and percid community structure. *J. Fish. Res. Board Canada* 34:1952–1958.

———. 1989. Current approaches to multispecies analyses of marine fisheries. *Can. J. Fish. Aquat. Sci.* 46:528–534.

———. 1997. The Laurentian Great Lakes experience: A prognosis for the fisheries of Atlantic Canada. *Can. J. Fish. Aquat. Sci.* 54:1190–1197.

Koestler, A. and J. R. Smythies. 1969. *Beyond Reductionism: New Perspectives in the Life Sciences.* Boston: Beacon Press.

Kriksunov, Ye. A. and M. L. Shatunovsky. 1979. Some questions of population structure variability in the smelt, *Osmerus eperlanus. J. Ichthyology* 19: 48–55.

Krohn, M. M. and D. Boisclair. 1994. Use of a stereo-video method to estimate the energy expenditure of free-swimming fish. *Can J. Fish. Aquat. Sci.* 51: 1119–1127.

Krohn, M. M. 1999. Growth and bioenergetics of Northern Cod (*Gadus morhua*). Ph.D. thesis, Biology Department, Dalhousie University, Halifax, N.S.

Lawrie, A. H. and J. F. Rahrer. 1972. Lake Superior: Effects of exploitation and introductions on the salmonid community. *J. Fish. Res. Board Canada* 29: 765–776.

Leach, J. H., L. M. Dickie, B. J. Shuter, U. Borgmann, J. Hyman, and W. Lysak. 1987. A review of methods for prediction of potential fish production with application to the Great Lakes and Lake Winnipeg. *Can. J. Fish. Aquat. Sci.* 44(Suppl. II):471–485.

LeShan, L. and H. Margenau. 1982. *Einstein's Space and Van Gogh's Sky.* New York: Collier Books.

Levine, S. 1980. Several measures of trophic structure applicable to complex food webs. *J. Theor. Biol.* 83:195–207.

Levins, R. 1968. *Evolution in Changing Environments: Some Theoretical Explorations.* Princeton, N.J.: Princeton University Press.

Li, W. K. W. and P. M. Dickie. 1985. Growth of bacteria in seawater filtered through 0.2 micrometer Nucleopore membranes: Implications for dilution experiments. *Mar. Ecol. Progr. Series* 26:245–252.

Lindeman, R. L. 1942. The trophic-dynamic aspect of ecology. *Ecology* 23:399–418.

Loder, J. W. and D. A. Greenberg. 1986. Predicted positions of tidal fronts in the Gulf of Maine region. *Continental Shelf Res.* 6:397–414.

Longhurst, A. C. 1967. Vertical distribution of zooplankton in relation to the Pacific oxygen minimum. *Deep Sea Res.* 14:51–63.

Lotka, A. J. 1956. *Elements of Mathematical Biology.* New York: Dover. [Reprint of *Elements of Physical Biology,* Williams & Watkins, 1924.]

Ludwig, D. and C. J. Walters. 1985. Are age-structured models appropriate for catch-effort data? *Can. J. Fish. Aquat. Sci.* 42:1066–1072.

MacArthur, R. H. 1957. On the relative abundance of bird species. *Proc. Nat. Acad. Sci. Wash.* 43:293–295.

———. 1960. On the relative abundance of species. *Amer. Naturalist* 94: 25–36.

MacCall, A. D. 1976. Density dependance of catchability coefficient in the Californian Pacific sardine, *Sardinops sagax caerula,* purse-seine fishery. *Mar. Res. Comm. Calif. Coop. Oceanic Fish. Invest. Rep.* 18:136–148.

MacPherson, E. and A. Gordoa. 1996. Biomass spectra in benthic fish assemblages in the Benguela system. *Mar. Ecol. Progr. Series* 138:27–32.

Mann, K. H. 1969. The dynamics of aquatic ecosystems. *Adv. Ecol. Res.* 6:1–81.

Mann, K. H. and J. R. N. Lazier. 1991. *Dynamics of Marine Ecosystems: Biological-Physical Interactions in the Oceans.* Oxford: Blackwell.

Margalef, R. 1968. *Perspectives in Ecological Theory.* Chicago: University of Chicago Press.

Martin, N. V. 1966. The significance of food habits in the biology, exploitation, and management of Algonquin Park, Ontario, Lake Trout. *Trans. Amer. Fish. Soc.* 95:415–422.

————. 1970. Long-term effects of diet on the biology of the lake trout and the fishery in Lake Opeongo, Ontario. *J. Fish. Res. Board Canada* 27:125–146.

Mayr, E. 1997. *This Is Biology: The Science of the Living World.* Cambridge, Mass: Harvard University Press.

McLaren, I. A . 1966. Physical and chemical characteristics of Ogac Lake, a landlocked fiord on Baffin Island. *J. Fish. Res. Board Canada* 24:981–1015.

————. 1969a. Primary production and nutrients in Ogac Lake, a landlocked fiord on Baffin Island. *J. Fish. Res. Board Canada* 26:1561–1576.

————. 1969b. Population and production ecology of zooplankton in Ogac Lake, a landlocked fiord on Baffin Island. *J. Fish. Res. Board Canada* 26:1485–1559.

McQuinn, I. H. 1997. Metapopulations and the Atlantic herring. *Rev. Fish Biol. Fish.* 7:297–329.

Mills, E. L. and R. O. Fournier. 1979. Fish production and the marine ecosystems of the Scotian Shelf, Eastern Canada. *Mar. Biol.* 54:101–108.

Minns, C. K., E. S. Millard, J. M. Cooley, M. G. Johnson, D. A. Hurley, K. H. Nicholls, G. W. Robinson, G. E. Owen, and A. Crowder. 1987. Production and biomass size-spectra in the Bay of Quinte, a eutrophic ecosystem. *Can. J. Fish. Aquat. Sci.* 44(Suppl. 2):148–155.

Moloney, C. L. and J. G. Field. 1985. Use of particle-size data to predict potential pelagic-fish yield of some South African areas. *S. Afr. J. Mar. Sci.* 3:119–128.

Moloney, C. L., J. G. Field, and M. I. Lucas. 1991. The size-based dynamics of plankton food webs. II. Simulations of three contrasting southern Benguela food webs. *J. Plankton Res.* 13:1039–1092.

Myers, R. A. 1998. When do environment-recruit correlations work? *Rev. Fish Biol. Fish.* 8:285–305.

Myers, R. A., N. J. Barrowman, J. A. Hutchings, and A. A. Rosenberg. 1995. Population dynamics of exploited fish stocks at low population levels. *Science* 269:1106–1108.

Myers, R. A., J. A. Hutchings, and N. J. Barrowman. 1996. Hypotheses for the decline of cod in the North Atlantic. *Mar. Ecol. Progr. Series* 138:293–308.

———. 1997. Why do fish stocks collapse? The example of cod in Atlantic Canada. *Ecol. Applic.* 7:91–106.

Neave, F. 1954. Principles affecting the size of pink and chum salmon populations in British Columbia. *J. Fish. Res. Board Canada* 9:450–491.

Nelson, J. A., Y. Tang, and R. G. Boutilier. 1994. Differences in exercise physiology between two Atlantic cod (*Gadus morhua*) populations from different environments. *Physiol. Zool.* 67:330–354.

Nikolsky, G. V. 1963. *The Ecology of Fishes*. Translated by L. Birkett. London: Academic Press.

Nixon, S. W., C. A. Oviatt, J. Frithsen, and B. Sullivan. 1986. Nutrients and the productivity of estuarine and coastal marine ecosystems. *J. Limnol. Soc. S. Afr.* 12:43–71.

Odum, E. P,. 1969. The strategy of ecosystem development. *Science* 164:262–270.

Odum, E. P. and A. E. Smalley. 1959. Comparison of population energy flow of a herbivorous and a deposit-feeding invertebrate in a salt marsh ecosystem. *Proc. Nat. Acad. Sci. Wash.* 45:617–622.

O'Neill, R. V., D. L. DeAngelis, J. B. Waide, and T. F. H. Allen. 1986. *A Hierarchical Concept of Ecosystems*. Princeton, N.J.: Princeton University Press.

Oppenheimer, R. 1958. The growth of science and the structure of culture. *Daedalus (Proc. Amer. Acad. Arts and Sci.)* 87(1):67–76.

Packard, T. T. 1985. Measurement of electron transport activity in microplankton. In H. W. Jannasch and P. J. LeB. Williams, eds., *Advances in Aquatic Microbiology*, vol. 3, pp. 207–261. San Diego, CA: Academic Press.

Paley, W. 1802. *Natural Theology*. London: Rivington.

Paloheimo, J. E. and L. M. Dickie. 1964. Abundance and fishing success. *Rapp. Proc.-verb. Reun. Cons. Int. Explor. Mer* 155:152–163.

———. 1965. Food and growth of fishes I. A growth curve derived from experimental data. *J. Fish. Res. Board Canada* 22:521–542.

———. 1966a. Food and growth of fishes. II. Effects of food and temperature on the relation between metabolism and body-weight. *J. Fish. Res. Board Canada* 23:869–908.

———. 1966b. Food and growth of fishes. III. Relations among food, growth and body efficiency. *J. Fish. Res. Board Canada* 23:1209–1248.

Parsons, T. R. 1969. The use of particle size spectra in determining the structure of a plankton community. *J. Oceanogr. Soc. Japan* 25:172–181.

Parsons, T. R. and R. J. LeBrasseur. 1970. The availability of food to different trophic levels in the marine food chain. In J. H. Steele, ed., *Marine Food Chains*, pp. 325–343. Edinburgh: Oliver and Boyd.

Parsons, T. R., R. J. LeBrasseur, and J. D. Fulton. 1967. Some observations on the dependance of zooplankton grazing on the cell size and concentration of phytoplankton blooms. *J. Oceanogr. Soc. Japan* 23:10–17.

Parsons, T. R., M. Takahashi, and B. Hargrave. 1984. *Biological Oceanographic Processes.* 3d ed. Oxford: Pergamon Press.

Patriquin, D. G. 1966. Biology of *Gadus morhua* in Ogac Lake, a landlocked fiord on Baffin Island. *J. Fish. Res. Board Canada* 24:2573–2594.

Paulik, G. J. 1971. Anchovies, birds and fishermen in the Peru Current. In W. W. Murdock, ed., *Environment, Resources, Pollution, and Society,* pp. 156–185. Stamford, Conn.: Sinauer Assoc.

Pauly, D., V. Christensen, J. Dalsgaard, R. Froese, and F. Torres Jr. 1998. Fishing down food webs. *Science* 279:860–863.

Peters, R. H. 1983. *The Ecological Implications of Body Size.* Cambridge, Mass.: Cambridge University Press.

Petrusewicz, K. and A. Macfadyen. 1970. *Productivity of Terrestrial Animals; Principles and Methods.* International Biological Program Handbook No. 13. Oxford: Blackwell.

Platt, T. 1985. Structure of the marine ecosystem: Its allometric basis. *Can. Bull. Fish. Aquat. Sci. Ottawa* 213:55–64.

Platt, T. and K. Denman. 1975. Spectral analysis in ecology. *Ann. Rev. Ecol. Syst.* 6:189–210.

———. 1977. Organization in the pelagic ecosystem. *Helgol. wiss. Meeres.* 30:575–581.

———. 1978. The structure of pelagic marine ecosystems. *Rapp. Proc.-verb. Réun. Cons. Int. Explor. Mer* 173:60–65.

Platt, T., M. Lewis, and R. Geider. 1984. Thermodynamics of the pelagic ecosystem, elementary closure conditions for biological production in the open ocean. In M. R. J. Fasham, ed., *Flows of Energy and Materials in Marine Ecosystems,* pp. 49–84. London: Plenum.

Platt, T. and W. K. W. Li. 1986. Photosynthetic picoplankton. *Can. Bull. Fish. Aquat. Sci. Ottawa* 214:1–583.

Platt, T. and W. Silvert. 1981. Ecology, physiology, allometry, and dimensionality. *J. Theor. Biol.* 93:855–860.

Pogson, G. H., K. A. Mesa, and R. G. Boutilier. 1995. Genetic population structure and gene flow in the Atlantic cod *Gadus morhua:* A comparison of nuclear and RFLP loci. *Genetics* 139:375–385.

Pope, J. G., T. K. Stokes, S. A. Murawski, and S. I. Idoine. 1988. A comparison of fish size-composition in the North Sea and on Georges Bank. In W. Wolfe, C.-J. Soeder, and F. R. Drepper, eds., *Ecodynamics: Contributions to Theoretical Ecology,* pp. 146–152. Berlin: Springer-Verlag.

Prigogine, I. and I. Stengers. 1984. *Order Out of Chaos: Man's New Dialogue with Nature.* Toronto: Bantam Books.

Quiñones, R. B. 1992. Size-distribution of planktonic biomass and metabolic activity in the pelagic system. Ph.D thesis, Faculty of Graduate Studies, Dalhousie University, Halifax, N.S.

Quiñones, R. B., J. M. Blanco, F. F. Echevarría, M. L. Fernández-Puelles, J. Gilabert, V. Rodríguez, and L. Valdes. 1994. Metabolic size spectra at a frontal station in the Alboran Sea. *Scientia Marina* 58:53–58.

Ramsay, P. M., S. D. Rundle, M. J. Attrill, M. G. Uttley, P. R. Williams, P. S. Elsmere, and A. Abada. 1997. A rapid method for estimating biomass size spectra of benthic metazoan communities. *Can. J. Fish. Aquat. Sci.* 54:1716–1724.

Rashevsky, N. 1959. Some remarks on the mathematical theory of nutrition of fishes. *Bull. Math. Biol.* 21:161–183.

Regier, H. A. 1973. Sequence of exploitation of stocks in multispecies fisheries in the Laurentian Great Lakes. *J. Fish. Res. Board Canada* 30:1992–1999.

Reidy, S. P., J. A. Nelson, Y. Tang, and S. R. Kerr. 1995. Post-exercise metabolic rate in Atlantic cod and its dependance upon the method of exhaustion. *J. Fish Biol.* 47:377–386.

Ricker, W. E. 1940. Relation of "catch per unit effort" to abundance and rate of exploitation. *J. Fish. Res. Board Canada* 5:43–70.

———. 1954. Stock and recruitment. *J. Fish. Res. Board Canada* 11:559–623.

———. 1969. Food from the sea. In Preston Cloud, chairman, *Resources and Man,* pp. 87–108. San Francisco: W. H. Freeman.

———. 1975. *Computation and Interpretation of Biological Statistics of Fish Populations.* Bulletin of the Fisheries Research Board of Canada, no. 191. Ottawa.

Riley, G. A. 1963. Theory of food-chain relations in the sea. In M. N. Hill, ed., *The Sea,* vol. 2, pp. 438–463. New York: Interscience.

Ritchie, M. E. and H. Olff. 1999. Spatial scaling laws yield a synthetic theory of biodiversity. *Nature* 400:557–560.

Roa, R. and R. A. Quiñones. 1998. Theoretical analysis of the relationship between production per unit biomass and animal body size. *Oikos* 81:161–167.

Rodríguez, J., F. Echevarría, and F. Jimenez-Gomez. 1990. Physiological and ecological scalings of body-size in an oligotrophic high mountain lake (LaCaldera, Sierra Nevada, Spain). *J. Plank. Res.* 12:593–599.

Rodríguez, J., F. Jimeniz, B. Bautista, and V. Rodríguez. 1987. Planktonic biomass spectra dynamics during a winter production pulse in Mediterranean coastal waters. *J. Plank. Res.* 9:1183–1194.

Rodríguez, J. and M. M. Mullin. 1986. Relations between biomass and body-weight of plankton in a steady state oceanic system. *Limnol. Oceanogr.* 21:361–370.

Rojo, C. and J. Rodríguez. 1994. Seasonal variability of phytoplankton size structure in a hypertrophic lake. *J. Plank. Res.* 16:317–335.

Rose, G. A. and W. C. Leggett. 1991. Effects of biomass-range interactions on catchability of migratory demersal fish by mobile fisheries: An example of Atlantic cod (*Gadus morhua*). *Can. J. Fish. Aquat. Sci.* 48:843–848.

Rosen, R. 1972. On the decomposition of a dynamical system into non-interacting subsystems. *Bull. Math. Biophys.* 34:337–341.

———. 1978. *Fundamentals of Measurement and Representation of Natural Systems.* New York: Elsevier.

———. 1991. *Life Itself.* New York: Columbia University Press.

———. 1999. *Essays on Life Itself.* New York: Columbia University Press.

Ruzzante, D. E., C. T. Taggart, and D. Cook. 1996. Spatial and temporal variation in the genetic composition of a larval cod (*Gadus morhua*) aggregation: Cohort contribution and genetic stability. *Can. J. Fish. Aquat. Sci.* 53:2695–2705.

Ruzzante, D. E., C. T. Taggart, D. Cook, and S. Goddard. 1996. Genetic differentiation between inshore and offshore Atlantic cod (*Gadus morhua*) off Newfoundland: Microsatellite DNA variation and antifreeze level. *Can. J. Fish. Aquat. Sci.* 53:634–645.

Ryder, R. A. and S. R. Kerr. 1990. Harmonic communities in aquatic ecosystems: A management perspective. In W. L. T. van Densen, B. Steinmetz, and R. H. Hughes, eds., *Management of Freshwater Fisheries,* pp. 594–623. Wageningen: Pudoc.

Ryder, R. A., S. R. Kerr, W. W. Taylor, and P. A. Larkin. 1981. Community consequences of fish stock diversity. *Can. J. Fish. Aquat. Sci.* 38:1856–1866.

Sameoto, D., N. Cochrane, and A. Herman. 1993. Convergence of acoustic, optical and net-catch estimates of euphausiid abundance: Use of artificial light to reduce net avoidance. *Can. J. Fish. Aquat. Sci.* 50:334–346.

Sameoto, D. D., L. O. Jaroszynski, and W. B. Fraser. 1977. A multiple opening and closing plankton sampler based on the MOCNESS and N.I.O. nets. *J. Fish. Res. Board Canada* 34:1230–1235.

Saville, A. 1979. Discussion and conclusions of the symposium on the biological basis of pelagic fish stock management. *Rapp. Proc.-verb. Réun. Cons. Int. Explor. Mer* 177:513–517.

Schaefer, M. B. 1967. Fishery dynamics and present status of the yellowfin tuna population of the eastern Pacific Ocean. *Bull. Inter-Amer. Trop. Tuna Comm.* 12:89–136.

Schaefer, M. B. and R. J. H. Beverton. 1963. Fishery dynamics—their analysis and interpretation. In M. N. Hill, ed., *The Sea,* vol. 2, pp. 464–483. New York: Interscience.

Schmidt-Nielsen, K. 1984. *Scaling: Why Is Animal Size So Important?* Cambridge: Cambridge University Press.

Schoener, T. W. 1974. Resource partitioning in ecological communities. *Science* 185:27–39.

Schwinghamer, P. 1981a. Extraction of living meiofauna from marine sediments by centrifugation in a silica sol-sorbitol mixture. *Can. J. Fish. Aquat. Sci.* 38:476–478.

———. 1981b. Characteristic size distributions of integral benthic communities. *Can. J. Fish. Aquat. Sci.* 38:1255–1263.

———. 1983. Generating ecological hypotheses from biomass spectra using causal analysis: A benthic example. *Mar. Ecol. Progr. Series* 13:151–166.

———. 1986. Observations of size-structure and pelagic coupling of some shelf and abyssal benthic communities. In P. E. Gibbs, ed., *Proceedings of the 19th European Marine Biology Symposium,* pp. 347–359. Cambridge: Cambridge University Press.

Schwinghamer, P., J. Y. Guigné, and W. C. Siu. 1996. Quantifying the impact of trawling on benthic habitat structure using high resolution acoustics and chaos theory. *Can. J. Fish. Aquat. Sci.* 53:288–296.

Schwinghamer, P., B. Hargrave, D. Peer, and C. M. Hawkins. 1986. Partitioning of production and respiration among size groups of organisms in an intertidal benthic community. *Mar. Ecol. Progr. Series* 31:131–142.

Sheldon, R. W. and T. R. Parsons. 1967. A continuous size spectrum for particulate matter in the sea. *J. Fish. Res. Board Canada* 24:909–915.

Sheldon, R. W., A. Prakash, and W. H. Sutcliffe Jr. 1972. The size distribution of particles in the ocean. *Limnol. Oceanogr.* 17:323–340.

Sheldon, R. W., W. H. Sutcliffe Jr., and M. A. Paranjape. 1977. Structure of pelagic food chain and relationship between plankton and fish production. *J. Fish. Res. Board Canada* 34:2344–2353.

Sheldon, R. W., W. H. Sutcliffe Jr., and A. Prakash. 1973. The production of particles in the surface waters of the ocean with particular reference to the Sargasso Sea. *Limnol. Oceanogr.* 18:719–733.

Sieburth, J. M., V. Smetacek, and J. Lenz. 1978. Pelagic ecosystem structure: Heterotrophic compartments of plankton and their relationship to plankton size fractions. *Limnol. Oceanogr.* 23:1256–1263.

Silvert, W. and T. Platt. 1978. Energy flux in the pelagic ecosystem: A time-dependent equation. *Limnol. Oceanogr.* 23:813–816.

———. 1980. Dynamic energy-flow model of the particle size-distribution in pelagic ecosystems. In W. C. Kerfoot, ed., *Evolution and Ecology of Zooplankton Communities,* vol. 3, pp. 754–763. Special Symposium of the American Society of Limnology and Oceanography. Hanover, N.H.: University Press of New England.

Sissenwine, M. P. 1984. Why do fish populations vary? In R. M. May, ed., *Exploitation of Marine Communities,* pp. 59–94. Berlin: Springer-Verlag.

———. 1986. Perturbation of a predator-controlled continental shelf ecosystem. In K. Sherman and L. M. Alexander, eds., *Variability and Management of Large Marine ecosystems,* pp. 55–85. AAAS Selected Symposium 99. Boulder, Colo.: Westview Press.

Sissenwine, M. P., B. E. Brown, J. E. Palmer, and R. J. Essig. 1982. Empirical examination of population interactions for the fishery resources off the northeastern USA. In M. C. Mercer, ed., *Multispecies Approaches to Fisheries Management Advice,* pp. 82–94. Canadian Special Publications of Fisheries and Aquatic Science, no. 59. Ottawa.

Skud, B. E. 1975. Revised estimates of halibut abundance and the Thompson-Burkenroad debate. International Pacific Halibut Commission, Scientific Report no. 56. Seattle.

———. 1982. Dominance in fishes: The relation between environment and abundance. *Science* 216:144–149.

Slobodkin, L. B. 1963. *Growth and Regulation of Animal Populations.* New York: Holt, Rinehart & Winston.

Smith F. E. 1976. Ecosystems and evolution. *Bull. Ecol. Soc. Amer.,* Spring, 2–6.

Smith, P. C. 1989. Circulation and dispersion on Browns Bank. *Can. J. Fish. Aquat. Sci.* 46:539–559.

Sparholt, H. 1985. The population, survival, growth, reproduction and food of arctic charr, *Salvelinus alpinus* (L.), in four unexploited lakes in Greenland. *J. Fish Biol.* 26:313–330.

Spencer, P. D. and J. S. Collie. 1997. Effect of nonlinear predation rates on rebuilding the Georges Bank haddock (*Melanogrammus aeglefinus*) stock. *Can. J. Fish. Aquat. Sci.* 54:2920–2929.

Sprules, W. G. 1980. Zoogeographic patterns in the size structure of zooplankton communities, with possible applications to lake ecosystem modeling and management. In W. C. Kerfoot, ed., *Evolution and Ecology of Zooplankton*

*Communities,* vol. 3, pp. 642–656. Special Symposium of the American Society of Limnology and Oceanography. Hanover, N.H.: University Press of New England.

———. 1984. Towards an optimal classification of zooplankton for lake ecosystem studies. *Verh. Int. Verein. Limnol.* 22:320–325.

Sprules, W. G., S. B. Brandt, D. J. Stewart, M. Munawar, E. H. Jin, and J. Love. 1991. Biomass size spectrum of the Lake Michigan pelagic food web. *Can. J. Fish. Aquat. Sci.* 48:105–115.

Sprules, W. G., J. M. Casselman, and B. J. Shuter. 1983. Size distributions of pelagic particles in lakes. *Can. J. Fish. Aquat. Sci.* 40:1761–1769.

Sprules, W. G. and A. P. Goyke. 1994. Size-based structure and production in the pelagia of Lakes Ontario and Michigan. *Can. J. Fish. Aquat. Sci.* 51:2603–2611.

Sprules, W. G. and M. Munawar. 1986. Plankton size spectra in relation to ecosystem productivity, size and perturbation. *Can. J. Fish. Aquat. Sci.* 43:1789–1794.

Sprules, W. G. and J. D. Stockwell. 1995. Size-based biomass and production models in the St. Lawrence Great Lakes. *ICES J. Mar. Sci.* 52:705–710.

Steele, J. H. 1974. *The Structure of Marine Ecosystems.* Oxford: Blackwell.

———. 1978. Some comments on plankton patches. In J. H. Steele, ed., *Spatial Pattern in Plankton Communities.* New York: Plenum Press.

———. 1991. Can ecological theory cross the land-sea boundary? *J. Theor. Biol.* 153:425–436.

Steele, J. H. and E. W. Henderson. 1984. Modeling long-term fluctuations in fish stocks. *Science* 224:985–987.

Strayer, D. 1986. The size structure of a lacustrine zoobenthic community. *Oecologia* 69:513–516.

Sutcliffe, W. H., Jr., K. Drinkwater, and B. S. Muir. 1977. Correlations of fish catch and environmental factors in the Gulf of Maine. *J. Fish. Res. Board Canada* 34:19–30.

Swain, D. P. and A. F. Sinclair. 1994. Fish distribution and catchability: What is the appropriate measure of distribution? *Can. J. Fish. Aquat. Sci.* 51:1046–1054.

Thiebaux, M. L. and L. M. Dickie. 1992. Models of aquatic biomass size spectra and the common structure of their solutions. *J. Theor. Biol.* 159:147–161.

———. 1993. Structure of the body-size spectrum of the biomass in aquatic ecosystems: A consequence of allometry in predator-prey interactions. *Can. J. Fish. Aquat. Sci.* 50:1308–1317.

Thompson, D'Arcy W. 1917. *On Growth and Form.* Abridged edition 1961. Cambridge: The University Press.

Thompson, W. F. and F. H. Bell. 1934. *Biological Statistics of the Pacific Halibut Fishery (2) Effect of Change in Intensity upon Total Yield and Yield per Unit of Effort.* International Fisheries (Pacific Halibut) Commission, Report no. 8. Seattle.

Thompson, W. F., H. A. Dunlop, and F. H. Bell. 1931. *Biological Statistics of the Pacific Halibut Fishery (1) Changes in Yield of a Standardized Unit of Gear.* International Fisheries (Pacific Halibut) Commission, Report no. 6. Seattle.

Ulanowicz, R. E. 1997. *Ecology, the Ascendent Perspective.* New York: Columbia University Press.

Van der Meer, Jaap. 1998. Theoretical analysis of the relationship between production per unit biomass and animal body-size: A comment. *Oikos* 83(2): 331–332.

Vezina, A. F. 1985. Empirical relationships between predator and prey size among terrestrial vertebrate predators. *Oecologia* 67:555–565.

———. 1986. Body size and mass flow in freshwater plankton: Models and tests. *J. Plank. Res.* 8:939–956.

Vidondo, B., Y. T. Prairie, J. M. Blanco, and C. M. Duarte. 1997. Some aspects of the analysis of size spectra in aquatic ecology. *Limnol. Oceanogr.* 42:184–192.

Walters, C. J. 1986. *Adaptative Management of Renewable Resources.* New York: Macmillan.

Walters, C. J. and J. S. Collie. 1987. Is research on environmental factors useful to fisheries management? *Can. J. Fish. Aquat. Sci.* 45:1848–1854.

Walters, C. J., J. R. A. Park, and J. F. Koonce. 1980. Dynamic models of lake ecosystems. In E. D. LeCren and R. H. Lowe-McConnell, ed., *The Functioning of Freshwater Ecosystems,* pp. 455–479. Cambridge: Cambridge University Press.

Ware, D. M. 1980. Bioenergetics of stock and recruitment. *Can. J. Fish. Aquat. Sci.* 37:1012–1024.

Warwick, R. M. 1984. Species size distributions in marine benthic communities. *Oecologia* (Berlin) 61:32–41.

Waters, T. F. 1969. The turnover ratio in production ecology of freshwater invertebrates. *Amer. Naturalist* 103:173–185.

West, G. B., J. H. Brown, and B. J. Enquist. 1997. A general model for the origin of allometric scaling laws in biology. *Science* 276:122–126.

Wilhelm, S. W. and C. A. Suttle. 1999. Viruses and nutrient cycles in the sea. *Bioscience* 47:781–788.

Wilson, J. A and L. M. Dickie. 1995. Parametric management of fisheries: An ecosystem-social approach. In S. Hanna and M. Munasinghe, eds., *Prop-*

*erty Rights in a Social and Ecological Context: Case Studies and Design Applications,* pp. 153–166. Washington, D.C.: Beijer International Institute of Ecological Economics and the World Bank.

Wilson, J. A., J. French, P. Kleban, S. McKay, and R. Townsend. 1991. Chaotic dynamics in a multiple species fishery: A model of community predation. *Ecol. Modelling* 58:303–322.

Winters, G. H. and J. P. Wheeler. 1985. Interaction between stock area, stock abundance, and catchability coefficient. *Can. J. Fish. Aquat. Sci.* 42:989–998.

Witek, Z. 1986. Seasonal changes in the composition and size structure of plankton in the near-shore zone of the Gulf of Gdansk. *Ophelia* 4:287–298.

Witek, Z. and A. Krajewska-Soltys. 1989. Some examples of the epipelagic plankton size structure in high latitude oceans. *J. Plank. Res.* 11:1143–1155.

Zaret, T. M. and A. S. Rand. 1971. Competition in tropical stream fishes: Support for the competitive exclusion principle. *Ecology* 52:336–342.

Zhou, M. and M. E. Huntley. 1997. Population dynamics theory of plankton based on biomass spectra. *Mar. Ecol. Progr. Series* 159:61–73.

Zukav, Gary. 1980. *The Dancing Wu Li Masters: An Overview of the New Physics.* New York: Bantam Books.

# Index of First Authors

# SUBJECT INDEX

abundance
  of food, 130, 236
  of groups of species, 37, 40, 144, 196, 246
  of single species or stocks, 16, 49, 69, 176, 179, 190, 192, 193, 207, 210, 212, 213, 215, 217–220, 222–225, 228, 229, 247, 256, 260, 263, 284
  relation to density, 82–85, 175, 218, 221
adaptation, 24, 64, 112, 131, 135, 143, 144, 153, 188, 198, 206
allometric coefficient or parameter ($\gamma$), 125, 126, 164, 236
allometric relations, 28, 33, 40, 58, 79, 91, 160, 232, 233, 274, 276. *See also* physiological scaling (allometric coefficient of)
allometric studies, 35, 53, 69, 76, 94, 96, 97, 99, 100, 133, 196, 216, 235. *See also* conversion coefficient (allometric relations of)
analytical spectrum (internal or secondary structuring), 15, 27, 39, 46, 72, 74, 76, 85, 111, 114, 117, 121, 124–134, 156, 158, 181, 201–203, 212, 216
anchovetta, 248
aquatic communities. *See* communities

biomass body-size spectrum, 1–3, 7, 9–11, 13, 14, 19, 63–70, 156–159, 173, 176, 177, 187, 195, 197, 201, 209, 248–251, 257–259
  Model I, 103, 106, 107, 114, 115, 126
  Model II, 103, 106, 107, 109, 111, 114–117, 121, 122, 126, 127, 129, 133
  Model II (data fittings), 164, 212–216, 234
  normalized spectrum, 34, 36, 37, 40, 93, 240
  *See also* analytical spectrum; integral spectrum; spectrum(al) types
birds, 16, 25, 56, 142, 144, 147, 195, 196, 210
body-size distribution, 16, 46, 107, 195, 230
  biomass domes, 45, 63, 64, 66, 75, 92, 109, 115, 139, 158–160, 181, 182, 185, 187, 189, 190, 193, 198–200, 216, 275
  biomass subdomes, 77, 116, 117, 121–127, 130–132, 139, 181, 192, 230, 231, 237, 245
  discontinuities (gaps), 28, 29, 31, 63, 66, 108, 110, 168, 194, 195–197, 216
  prey availability, 130, 142, 236